Our Hidden Reality

Revealing the Foundational Structure

Richard L. Marker

" the book
nobody will read "

dedicated to

my wife Gini

DOI : 10.6084/m9.figshare.29215250

ISBN : 979-8-218-71537-3

email : rlmarker@spaceandmatter.org

First Edition, First Printing (July, 2025)

Table of Contents

Preface

Seeing the *unseeable* lies beyond our reach. Visualizing the *unseeable* lies within our reach. Many consider this impossible. There are good reasons for skepticism. Our physical reality conditions us to expect conformance with it. How could we have the experience or the language to understand or describe our hidden reality? Researchers may try to understand our hidden reality by extending our perceived reality to ever smaller concepts or particles. Some look for hidden dimensions. Finding a solution remains elusive.

Even an open-minded reader will find concepts in this book that run counter to their beliefs. It takes time to digest these thoughts and incorporate them into your world view. You may wonder if the reader needs to be a philosopher or physicist. Philosophers and physicists understand many deeper implications. Contrariwise, they incorporate established views into their thinking.

Our journey assumes that a structure underlies space, and that the electron forms an integral part of the structure. The building blocks for the structure are neighboring space-like elements that all behave the same. These building blocks lack any physical characteristics. At the foundational level there are no physical characteristics, there is no concept of time, distance, or geometry. Our task requires building a universe from this simple beginning.

The building blocks behave 100% causally mechanistically. The resulting structure could be called a discrete phase space where the phase difference between adjacent elements must shift by some fraction of a full circle. The fractional (i.e. rational) requirement leads to the quantum packet nature of interactions. Early on we find the electron needs self-entanglement (a single particle entangled with itself) and spin. The structure of the electron helps us to understand its dual wave-particle nature.

This book contains the details of Kant's neumona, that which is (almost) unknowable. They are so close to unknowable that I doubt they will ever be found again. Conjecture provides the only path to understanding neumona. Explanations of neumona challenge the reader to explore thoughts that fall far from physical reality.

Surprisingly, Nature provides a relationship that moves all of this conjecture into something closer to fact. Someone who enjoys a challenge will find the details in Appendix II.

Part I
Meaning of Physical Reality

Chapter 1
Different Realities

People may consider their perceived physical reality to be the only reality. I know the chair and the ball are real because I can feel the chair and sit in it; or I can throw the ball and expect its response to obey physical laws. Language builds around experiences.

1.1 Our perceived physical reality

We build much of our lives around our perception of reality. Others largely share our reality. Languages and physical laws reflect this reality. When Alice tumbles down the rabbit hole, her reality becomes distorted. We consider Alice's distorted reality to be a dream largely based on reality. Consider a rabbit hole in which hardly a fragment of our reality exists. Call it foundational reality.

Thinking about different realities presents challenges at best. The realities may differ so much that thinking about them falls outside of our ability to even conjecture.

Consider smaller and smaller regions of space. We know from quantum physics that some form of an unfamiliar reality exists at the quantum level. Now, consider regions of space so small that they are at the foundational level. What can we say or know about foundational reality? Can it even be understood in terms of our physical reality? These are important questions that can be answered only by knowing that which may be unknowable. Nevertheless, we must start somewhere to even stand a chance of understanding.

This book describes a journey that provides an understanding of *foundational reality*. The journey spanned nearly half a century and produced convincing results. The journey is about curiosity. ... about knowing the unknowable.

1.2 Evolving realities

Consider the everyday reality around us. We think everyone sees it somewhat similarly. It is not that clear. Look back to the time when people thought the world was flat. If you got too close to the edge of it you would fall off. Today we laugh at such thoughts. Back in time those thoughts were not laughable, they were people's reality.

Nicolaus Copernicus (1473-1543) changed prevailing thoughts about the universe. Previously, scientists considered the Earth as the center of the universe. Copernicus convinced others that it worked better to consider the Sun as the center of the Universe. Today we go beyond that view. People's thoughts and views of reality evolved.

Our views continue to evolve as we seek to better understand the physics of space. Einstein's *general relativity* (GR) reigned supreme for many decades. Today many work on the thankless task of patching GR with *dark matter* or cosmological constants. Mordehai Milgrom's modified gravity threw a monkey wrench into GR at great distances. GR already struggled at the quantum physics level. Now, it struggles at the cosmological level.

Looking inward at smaller and deeper scales, we apply our real life understandings and descriptions to quantum physics. Our inability to fully comprehend the strangeness of behaviors at that level demonstrates their departure from our everyday life. Now, go even deeper to *foundational physics.*

1.3 Language

With the flat-earth, our language allows us to describe an alternative globe-shaped earth. If we apply our language to quantum physics we find that it doesn't fit perfectly. Our language fits even less well at the foundational physics level. We try to fit our language to the birth of our universe and find an obscure notion called a singularity used to describe it. Our inability to describe different aspects of particular realities makes them even more challenging; or, perhaps impossible, to comprehend.

Our experiences and languages are based on what we perceive. How could we extend our language to go beyond our perceptions? Even an elaborate description of something must eventually be based on words or symbols that have a real life origin so we can understand what they mean. Keep this in mind as you read the various explanations of foundational physics.

1.4 Different realities, different laws

If we try to label two sets of behaviors as different realities, and they both have the same inherent laws or relationships then they would both seem to be the same reality. That is how we define different realities in this book. How much can the laws differ between realities? They could differ so much as to be incapable of being understood. If we can't describe the behaviors of a reality then it may lie beyond our ability to understand it.

Quantum physics provides an example of a reality that differs from our real world experiences. We describe an element of quantum physics as having a dual nature that exhibits some aspects of particles and some aspects of waves. This does not provide an answer; rather, it provides a paradox. The inexactness of measurement and the discrete nature of quantum packets both differ from our real world experiences. *Quantum reality* behaves quite differently from our real world reality.

1.5 Foundational reality

We consider *foundational reality* in this book. Foundational reality falls outside of our experience. It may even fall outside of the reach of our understanding. It could vary so dramatically as to be unfindable. Hopefully, finding the foundational reality will provide answers about both quantum reality and real world reality.

Some people refer to understanding the foundational reality as a *Theory of Everything (TOE)*. Personally, I find this terminology inaccurate. It seems to ignore the contributions made by many thousands of physicists and philosophers. A theory can't replace those contributions. I avoid using the label TOE.

Chapter 2
Understanding Foundational Reality

How can we start when trying to understand foundational reality? Equally importantly, how can we know when we have found that which we seek? All journeys such as this start with conjecture. To be meaningful the conjecture must lead somewhere.

The journey that provides the basis for this book initially started quite differently than described. The process started as a two year journey to disprove personal beliefs about the ether being structured. I felt that special relativity (SR) didn't fit well with my beliefs. I knew that mathematically SR was okay, but it seemed to remove a structured ether as a natural option. Working to disprove my beliefs did not go as planned. I then embarked on an unrealistic journey to find the foundational structure. Filled with over-confidence and naivety, I proceeded. That was almost half a century ago.

2.1 Michelson-Morley

The Michelson-Morley (M-M) experiment (1887)[a] failed to detect an ether. Today the M-M experiment is often erroneously taken to indicate the non-existence of an ether. This view limits or eliminates our ability to theorize a foundational structure.

George Francis FitzGerald's 1889 Letter[b] provided the thoughts needed to realize the M-M experiment was inconclusive.

> ... We know that electric forces are affected by the motion of the electrified bodies relative to the ether and it seems a not improbable supposition that the molecular forces are affected by the motion and that the size of the body alters consequently. ...

---

[a]Michelson, Albert A.; Morley, Edward W. (1887). *On the Relative Motion of the Earth and the Luminiferous Ether.* American Journal of Science 34 (203): 333-345. Bibcode:1887AmJS...34..333M.
`https://doi.org/10.2475/ajs.s3-34.203.333`

[b]FitzGerald, George F. (1889). "The Ether and the Earth's Atmosphere.", [Letters to the Editor],*Science*, vol. 13, p. 390.

The M-M experiment confounds researchers attempting to understand the foundational reality. SR introduces a view that seems contrary to an underlying structured ether. The *Lorentz transformation* proposed by Hendrik Lorentz in 1892 explained how the M-M experiment could be consistent with an ether. At the time Lorentz was unaware that FitzGerald had proposed a mechanism in 1889. The inability to detect an ether is now generally taken to mean an ether does not exist. The indetectibility of an ether differs from the nonexistence of an ether. The symmetrical views of SR seem counterintuitive to the existence of an ether. This moves people away from the ether concept.

The equations in Appendix I reconcile the views presented with an ether when compared to SR. In all equations we assume collinear motion with: observer A traveling at velocity u and observer B traveling at velocity v with respect to a fixed ether. Equation (1) gives the measured velocity, w, between the two observers.

Equation (8) and Equation (10) give the transformation for A's view of B and B's view of A, respectively. In both cases the same special relativistic transformation factor develops. The difference between the two equations lies in the real and the observational component of the transformation. If an observer is at rest, then their measurements are real and the observational component is unity. If an observer is in motion, then their measurements are distorted. In this case, the observational component is not unity.

Essentially, the effect of transformations is that whatever happens physically, also happens in the measuring device. So, differences seem to disappear. The measuring device can be a ruler; or, a M-M experiment. Later, we will find that this same effect happens with our 3-D interpretation of the foundational structure.

2.2 Mechanical view

A quote from Einstein's & Infeld's 1938 book[c] (p. 125), in the chapter titled: "The Decline of the Mechanical View":

> Science did not succeed in carrying out the mechanical program convincingly, and today no physicist believes in the possibility of its fulfillment.

This quote was a strong statement from a foremost leader in the physics community. I disagree and consider the foundational theory presented in this book to be 100% causal-mechanistic.

[c]Einstein, A., & Infeld, L. (1938). *The Evolution of Physics : the Growth of Ideas from Early Concepts to Relativity and Quanta.* In University Press eBooks (Vol. 4). `http://ci.nii.ac.jp/ncid/BA11638684`

2.3 Many hurdles

Understanding foundational reality requires both the ability to understand it and the language to express that understanding. Our physical reality could greatly limit our ability to understand foundational reality. It is possible that we cannot overcome the challenges. In this case, the journey to understand our foundational reality becomes a moot process.

Some of the many hurdles we face include :

- confidence – Understanding the foundational reality falls so far from our experiences that we need to remain confident that a solution will follow. This confidence will be tested many times.
- magic – Foundational reality cannot base itself on magic. Otherwise, comprehending it remains out of our reach. Allowing oneself to consider magic as an answer makes giving up easy.
- simplicity – At its heart, Nature must be simple. If it is not simple, then we are not at the beginning. The challenge faces us of using a simple beginning to build a complex universe.
- language – Foundational concepts and laws may be far removed from our physical reality. If so, our languages may be inadequate to communicate or understand the foundation.
- laws – How does one discover laws in a reality far removed from our physical reality? Even if we postulate the correct laws, how do we know they are correct?
- no physical characteristics – Any concept from physical reality that we bring into our thoughts about foundational reality limits our chances of successfully understanding it. At the foundational level we assume that time, distance, and geometry do not exist. This makes it especially hard to comprehend.
- confirmation – Our understanding of foundational reality needs confirmation to move it beyond speculation. How can Nature provide confirmation?
- utility – Even if we overcome the many hurdles, discovery of our foundational reality must provide utility. Otherwise, what was the point? It seems that utility should naturally emerge if our journey succeeds, so this may not be a hurdle.

Part II
Traveling into the Unknown

Chapter 3
The Impossible Journey

Finding the foundational reality seems like a fool's errand. This seemingly impossible journey provides the views shared in this book.

3.1 A practical beginning

Where do we start our journey of discovery? Practical requirements limit our choices. If these requirements preclude finding an answer then we may be destined to not find an answer.

Our practical beginning assumes the following:

- structure – A foundational structure of some type exists. Our task is to determine its behavior and how to build it.
- matter – The structure of foundational matter shares features of foundational space. If this holds, then we need find only one similar structure for both space and matter.
- simple elements – We start our structure with a very large number of identical simple elements. If the elements are not identical, then they are not simple elements. In order to differ they would need identifying characteristics, but we allow no characteristics. The elements can exhibit differing structural configurations, but they must all be the same and follow the same foundational laws.
- something & nothing – We refer to the simple elements as *somethings*. To consider many simple elements, we introduce an imaginary concept of *nothing* to separate the *somethings*. *Nothing* will do the heavy lifting that provides the structure.
- neighbors – *Somethings* between which events can occur. The concept of neighboring elements leads to structure.
- no physical characteristics – For foundational reality time, distance, and geometry do not exist. However, imaginary versions of these concepts help us understand the interaction laws and provide visualization.

3.2 Foundational granularity

The smallest element of the foundational structure can have no granularity. In order for granularity to exist there must be smaller elements. Consider an all encompassing foundation that consisted of everything the same with no elements whatsoever. There would be no interacting elements for which laws might be considered. There would be no reference points for motion to even have meaning. This constraint seems best addressed by having many starting foundational elements, all with no granularity or identifiable differences.

The complete lack of granularity means that no physical characteristics can be attributed to the foundational elements. The reason may not immediately be obvious. In order for there to be physical characteristics, there must be identifiable differences which require some granularity in order for differences to exist.

3.3 The first step

Before the first step we have a large(possibly infinite) number of *somethings*(foundational space elements) separated by an imaginary *nothing*. The *somethings* lack knowledge of each other and there is no coordinate system. Complete chaos exists because the *somethings* don't know what stationary even means. From this beginning we start the Impossible Journey to build our universe.

3.4 An unlikely causal event

Only one resolution to the chaos challenge presented itself after five years of intense thinking about *nothing*. Two of the *somethings* must have chanced to share an event. Is this even possible? Let's call the event a collision. In a reality where time, distance(size) and geometry don't exist a collision seems unlikely. Contrariwise, an infinite number of *somethings* combined with an infinite amount of imaginary time might facilitate an event with an infinitely small likelihood of occurring. Seeing no alternative, we proceed.

In our physical reality a collision may be elastic or inelastic, and the angle of the collision must be considered. In the foundational reality these concepts don't exist. We must guess. The simplest guess seemed a good place to start. We assume the collision to be complete(as opposed to partial). The collision event cancels opposing 'motions' in a causal averaging manner.

3.5 A periodic path

If we are to have a structure that endures, it seems likely that any 'motion' must be periodic and repeat itself. In mathematics there is a series of infinite sine and cosine functions called a Fourier series. For any path of periodic motion there exists a companion Fourier series. This is the extent of my knowledge about the series. The sine and cosine functions sound much like circles upon circles. I thought that higher order circles would eventually cancel out during collisions. The circle upon a circle leads to the Donut Chain Theory(DCT) in Appendix II. DCT has been included as an appendix in order to keep mathematics out of the conversation.

3.6 The foundational element

The foundational element(*something*) may be viewed as a *donut chain link*. The imaginary donut chain link has the features needed to evaluate interactions between neighboring foundational elements.

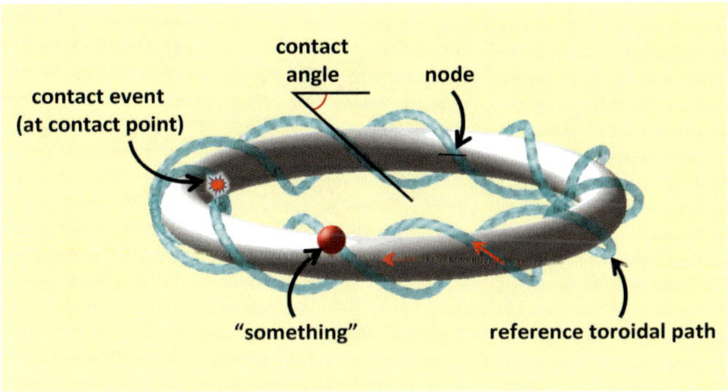

Figure 3.1: **Donut Chain Link**

The nodes shown in Figure 3.1. help us to find the chain segment structure of foundational space. The example reflects two revolutions around the main torus. The same *contact angle* develops if we spread the node count over a single toroidal revolution with a smaller poloidal radius. The single revolution version likely better represents the actual foundational phase relationships. (This may affect calculation extensions that explore quantum relationships.)

The donut chain link represents an *imaginary path*. The only foundationally real part of the donut chain link is labeled "*something*". *Something* is not a little round ball as shown.

Chapter 4
Chain Segments

A "connected" strand of chain links forms a *chain segment.* Later, we will find that the electron choreographs most of the details for the chain segments that form space and the electron itself.

4.1 Chain segments

Treating a space element as a donut chain link provides the details needed to determine the phase relationships between neighboring elements. Figure 4.1. shows a sample of donut chain linkage in normal space. Normal space chain segments are not twisted.

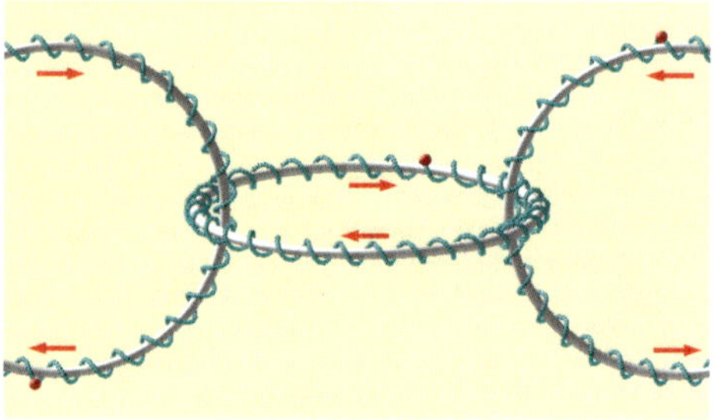

Figure 4.1: **Chain Segment Linkage**

The linkage shown in Figure 4.1. developed by assuming that an imaginary geometry could be treated as real. The imaginary geometric view exactly produces the phase relationships between adjacent elements of space. The imaginary geometric view does not extend to other aspects of the foundational structure.

4.2 Segment connecting links

Donut chain links that connect chain segments to each other orient their axes at phase angles of 0° or 180° . This provides uniform segment connections that help facilitate design symmetry and add stability. The imaginary geometry only holds for phasing it does not determine a geometry of the resulting structure.

4.3 The electron

The chain segment linkages shown in Figure 4.1. represent the connection between adjacent elements of space containing no electrons. The chain segments are untwisted. Each link is rotated 90° from its adjacent links. The chain links repeat phasing every four links. Chain segments with their ends 180° out of phase; repeat phasing after an even number of chain links not divisible by four. The segment link count convention includes a connecting link from only one of its ends.

Many node count solutions can work for the untwisted chain segments of normal space. However, the electron has one fewer links than the chain segments that form normal space. This missing link does not allow an exact solution for node counts. The impossibility of an exact solution for the twist angle leads to *self-entanglement*, spin, charge and mass.

There are reasons the electron has one fewer links:

- connected – The electron forms part of space. Its chain segment must somehow differ from a normal space chain segment.
- stability – One extra link in a chain segment can be disconnected as a neutrino to turn the chain segment into a normal space chain segment. This leads to instability. One fewer links requires decay of a baryonic particle to provide an extra link to turn the chain segment into a normal space chain segment. This leads to stability of chain segments with one fewer links.

There are consequences from one fewer links:

- twisted – Normal space is assumed to have the end connecting links of chain segments either in phase or 180° out of phase. An odd number of links requires twisting the chain segment. (Note: segment length convention only counts one end link.)
- uncertainty – There is no exact location for the electron. Point particle concepts do not apply at the foundational level. Consider a moving electron. Where is the electron located as it moves from one chain segment to the next?
- self-entanglement – The electron chain segment continually twists because there is no exact rational number for the twist angle present in the chain segment. As it twists the terminal connections move by one link to maintain phasing with other connected chain segments. This twisting and moving of the electron chain segment forms a connected strand in space that includes the electron. We call this *self-entanglement*.

4.4 The neutrino

The *neutrino* is a space element that has been disconnected from the structure. We might be tempted to call it a free space element not unlike the elements from which space chain segments initially form. This would be incorrect. The neutrino needs to be previously connected in order to subsequently synchronize and interact with space chain segments; or, matter chain segments. It needs to periodically interact with space in order to maintain its synchronization and to adjust to any time dilation. I don't know the speed at which a neutrino travels; or, its mass. Mass is an odd feature for the neutrino. Normally, particles that have mass are connected to the fabric of space. As such, their mass theoretically could be calculated. Neutrinos have no well defined connection with space, but they do periodically interact. It makes sense that the periodic interaction produces mass-like results, but they may be variable and not discernable in a definitive manner.

The periodic nature of neutrino interactions with space explains why they can travel through bodies unaffected. Neutrinos also are important balancing items in particle decay interactions. Some cosmic events produce huge numbers of neutrinos. Neutrino detection provides a valuable tool that helps us to understand the cosmos.

Neutrinos may help explain gamma-ray bursts (GRBs). GRB events produce large numbers of neutrinos. If GRBs provide a periodic connection with another universe, it would be a requirement that large numbers of neutrinos be produced when eliminating duplicate connections. The features of GRBs seem consistent with a connection between two universes.

I consider a universe to consist of semi-permanently connected chain segments. My view of a multiverse differs from parallel universes. There are many universes, but they cannot synchronize with each other in part due to different rates of the passage of time. Travel to another universe through a GRB would not be possible because the universes are not well synchronized.

Chapter 5

Entanglement / Spin

Entanglement and spin are closely related. These concepts need to be understood before meaningful calculations can be done. First, we look at normal vacuum space without an electron.

5.1 Vacuum space

Figure 5.1 shows a sample of chain segment connections for vacuum space. Note that it takes four donut chain links to return to the original phase. The sample shown has six links in each segment. The beginning and ending links in each chain segment are 180° out of phase. This image should not be considered geometric. It provides a convenient way of showing phase relationships.

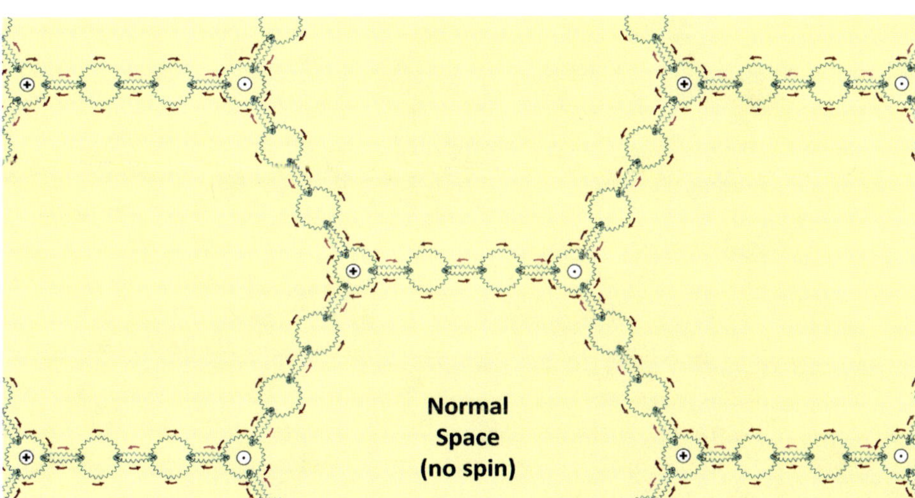

Figure 5.1: **Vacuum Space**

5.2 Self-entanglement / spin

Why do we introduce entanglement so early? It seems like a later concept. In our reality we experience electromagnetism and gravity, but not entanglement. Entanglement comes late in the game for reality. For a foundational search entanglement comes early. We must understand entanglement to continue our journey.

The need for spin arises from the need for a rational node count for each donut chain link combined with an irrational twist angle in the electron. The irrational twist angle can never be matched exactly. The angle mismatch causes a twisting motion in the electron chain segment. The twisting motion propagates through space chain segments until it reconnects with the other end of the electron chain segment to form a closed path.

Figure 5.2 shows a chain segment that is part of a self-entangled closed path. Note that when the chain segment rotates 90° it also moves to the right by one link. This cannot happen in isolation. A completed connection path forms. The entire closed path twists and moves. We call this *self-entanglement*. It is closely related to *spin*.

Figure 5.2: **Spin Details**

Take a moment to visually rotate the chain segment 90°. Notice that the rotating segment moves one link to the right. It does this by the side connections moving the opposite direction. Both ends of the rotating segment connect to a side connection and an adjacent segment that forms part of the closed-path self-entanglement strand.

5.3 Electron spin

Figure 5.3 shows a driver chain segment that generates spin together with a simplified closed-path strand of spin connected chain segments. Note the wind/unwind symbols. It seems likely that some of the side connections to the spin strand will wind and unwind as the spin strand rotates and moves forward one link.

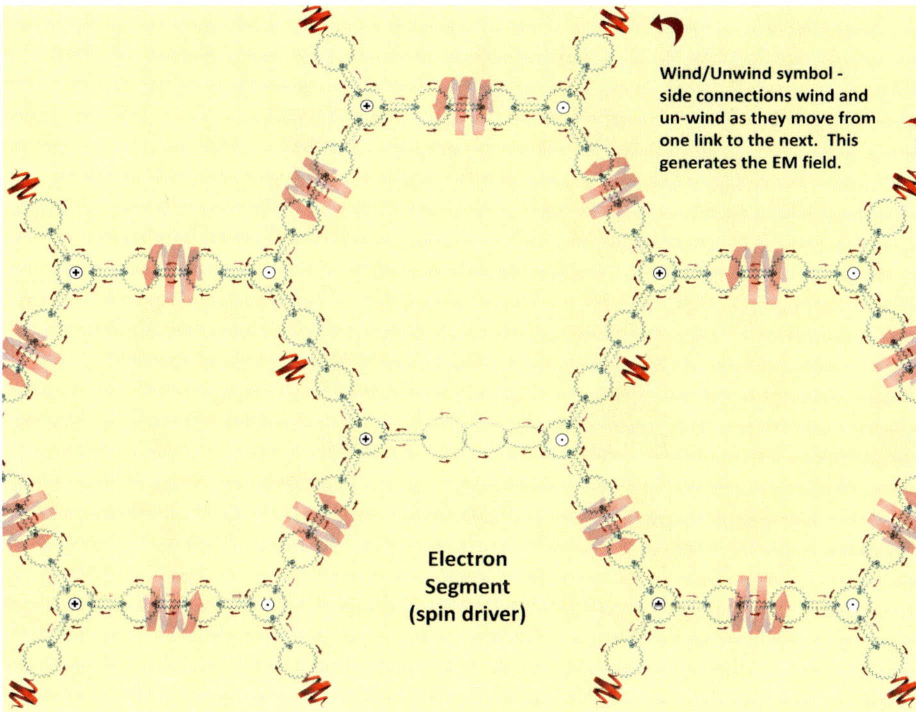

Figure 5.3: **Electron**

We will find that this closed-path spin strand concept needs to be understood in order to determine node counts. The light mauve spirals denote the continuous entanglement path. Note that the electron chain segment drives the spin and forms part of the path. The image shown does not represent an actual path. It merely illustrates concepts that apply to an actual path.

The ends of the electron chain segment are 180° out of phase. This may account for the 1/2 spin of the electron. The spin driver may also be a candidate to help explain the anomalous magnetic moment. The spin driver forms part of the spin connected closed-path chain strand. Its differences could affect the overall spin.

5.4 Photon spin

Figure 5.4 shows no driver chain segment for a photon. This is similar to the electron, but has an even number of chain segments instead of an odd number due to the missing driver segment.

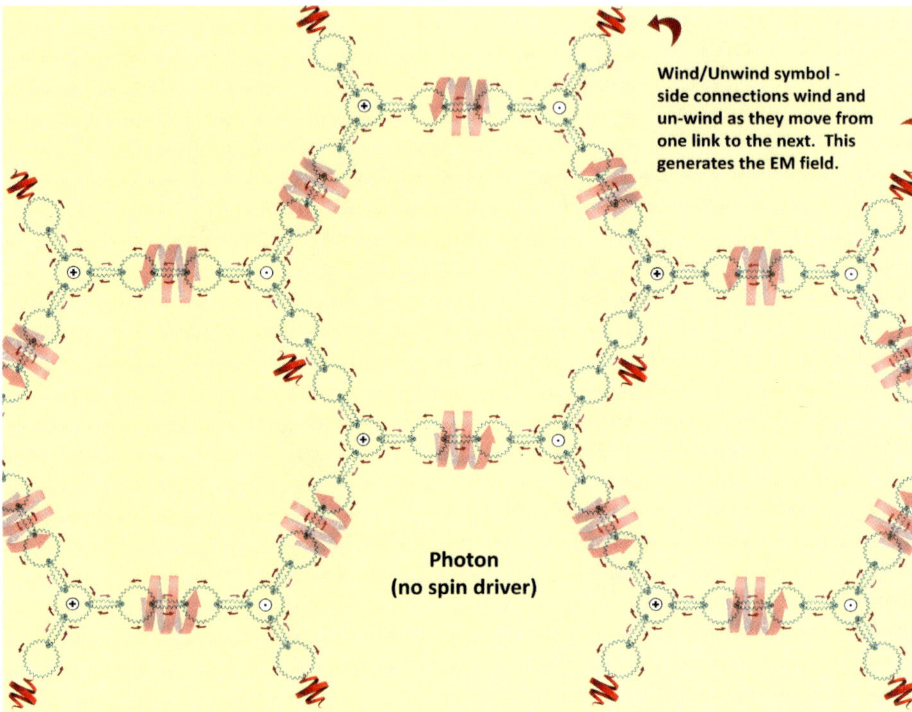

Figure 5.4: **Photon**

This image presents a crude representation of a photon. It does not explain the photon's propagation through space. The reader should not consider this image to be definitive.

The photon interacts with electromagnetic waves. The processes in which the photon participates add an important perspective to one's understanding of physics. However, the photon never fell into the category of 'need to know' in order to find the foundational structure. Hence, I never pursued an understanding.

5.5 Everyday examples

Entanglement provides the direct basis for many of our everyday applications. The behavior exhibited is usually ascribed to electromagnetism, but entanglement provides a more direct explanation.

Examples of entanglement in our everyday reality include:

- skin effect – When high frequency signals are sent over a conductor, the *skin effect* occurs. The signal is carried in the outer layer of the conductor. The electron in the conductor cannot physically move without a connected strand of space providing entanglement. With high frequencies (short wavelengths), only the outer layer of the conductor can provide the necessary entanglement connection. Inner layers of the conductor lack the time needed to establish the entanglement connection that occurs in space adjacent to the conductor.
- transformer – With transformers a precise ratio exists for the current in the output windings when compared with the current in the input windings. The ratio exactly equals the ratio of the number of windings in the output to the number of windings in the input. This effect is similar to the skin effect, except the output coils provide the return path for the signal.
- choke coil – A choke coil is like a transformer that is missing an output coil. Entanglement has no easy return path.
- coaxial cable connections – Did you ever watch a coaxial cable installer connect segments of coaxial cable? They take care to have the connections waterproof and electrically shielded. The waterproofing helps prevent corrosion. Why would they need to take such care with the shielding? It seems that a small signal leak would not lose much signal, but it does. A small leak provides an entanglement path for the signal. A small leak may be all that is needed to form a return entanglement path that allows much of the signal to escape.
- twisted pair ethernet cables – If you have worked with ethernet cables, you understand that each signal wire is twisted together with a companion ground wire. This pair of wires completes the return entanglement path. In this manner it keeps the signal better confined to the ethernet cable.
- atomic orbitals – Why do atomic orbitals adopt integer energy levels? The entanglement path length changes with incremental increases. Electron pairs can share an orbital because they have opposite spin which helps confine entanglement.

Chapter 6
Nodes

The word *nodes* occurs throughout this book. Nodes refers to the inward facing points of a spiral wrapped around the outer donut portion. Nodes are evenly distributed around the donut hole.

6.1 Node details

The example donut chain link shown in Figure 3.1. provides a visual that helps us understand the behavior between neighboring space elements. Three quantities need complementary values for an event to occur between space elements. The toroidal positions (main donut circle) must complement each other. The poloidal positions (outer circle) must be on the inside of each donut. Finally, these positions must occur at the same instant.

Nodes develop from imaginary concepts in foundational space. Curiously, the nodes and related donut interactions seem to exactly produce the behavior that occurs at the foundational level. This allows us to visualize foundational behavior.

Nodes may be considered as points uniformly distributed around the inside direction of the poloidal (outer) circle. They are event candidates, but must somehow connect with a similar event candidate at an exact 'location' and instant of time. Such exactness belies our experiences in physical reality. The rules of Nature at the foundational level differ from our physical reality. We see unusual behaviors at the quantum level. It should not surprise us to find even more unusual behaviors at the foundational level.

Donut Chain Theory (DCT) operates on exactness. A statistical aspect does not exist. Exactness indicates repeated interactive behavior. Nodes make this repetition possible. In order for donuts to have a stable node count, the ratio of poloidal (outer) revolutions to toroidal (main) revolutions must be rational if it is to be exact. It is the resolution of the paradox of achieving an irrational twist angle in the electron from rational revolution numbers in the nodes that leads to understanding foundational behavior.

Integer factors determine the node count. Otherwise, we could not have nodes occur both in stable toroidal and poloidal locations. Because of external chain segment relationships, we are able to constrain the factors needed in the electron's node count. *Self-entanglement* figures prominently in the external relationships.

Nodes may be viewed as teeth on a gear capable of transferring motion to another gear. Only for the instant that a tooth lines up exactly with a tooth in an adjacent gear will motion be transferred. The gears must be exactly synchronized and the displacement from any transferred motion must leave them synchronized, albeit with a different gear tooth. Motions are transferred in packets which are multiples of the amount needed for a one-tooth displacement.

The node count for a donut chain link includes a factor equal to the number of links in the chain segment. The inclusion of this factor allows the displacement at the ends of the chain segment to remain in the position needed for an event to occur. Self-entanglement of the electron requires that the node count for the electron include factors that are one greater and one less than the node count for normal space chain segments. This occurs because movement of the entangled strand changes the connecting link position such that one chain segment briefly adds a link while another loses a link. This add-a-link/subtract-a-link relationship *must* be included. For an example, see Equation (1) in Appendix II (page II.vii).

6.2 Twisted chain segment → spin

Consider the removal of a single link in a space chain segment (this is the electron). The end links of the modified chain segment can no longer be in phase or counterphase position without a twist. The shorter chain segment must be twisted by 90° in order to align with the normal phasing in space chain segments. Ignore 'distances' as they have no meaning in foundational reality.

We now have a problem that needs solving. Look at the *contact angle* in Figure 3.1. The tangent of the contact angle equals the poloidal node count divided by the toroidal node count. In normal space this has many rational solutions since the contact angle equals 45° . There is no exact rational node count solution that produces the irrational contact angle in the electron chain segment. How can we synchronize the chain links with an irrational contact angle? *Spin* seems to answer this problem.

Spin combined with a propagation delay along the connected spin strand(self-entanglement) could compensate for residual angle differences after we find a rational node solution that best produces the needed contact angle. Transmission of the needed propagation delay likely happens many times faster than the speed of light.

6.3 Guaranteed synchronization

The nodes in a donut chain link represent single tooth gears with numerous rational gear positions. The positions need to be rational in order for the pattern to repeat itself. There is one factor in the node count that provides a clue needed to solve relationships. For a chain segment that contains n links, an integer multiple(m) of n nodes in each link guarantees node synchronization in connecting links. Each link shifts node synchronization by m/n which leads to an overall shift of m nodes from one end of the chain segment to the other($m/n \cdot n = m$).

Guaranteed synchronization creates a larger number of nodes. This means finer gradations in the donut nodes. By using a factor equal to the number of donuts in a space chain segment, events can occur in smaller pieces while still achieving synchronization in the end connecting links. Solving for chain segment link counts and for the electron node count requires including this concept.

Guaranteed synchronization also plays a role in other events. For example, if we include p powers of 3 as a factor then synchronization across p space chain segments occurs. Remember that chain segment branching occurs at 120° phase intervals in the end connecting links.

6.4 Node implications of entanglement

Understanding the connected, closed-path strand of entanglement leads to inclusion of two node count factors that come from guaranteed synchronization. Chain segments in a path normally have n links. As the electron segment rotates 90° , the entanglement path must also rotate and move by one link. All of the entangled links do not move at the same instant. For a brief moment one chain segment participating in the entanglement will have $n + 1$ links while its next neighbor will have $n - 1$ links. Synchronization becomes far easier if each of the electron chain links has multiples of $(n + 1) \cdot (n - 1)$ links.

Finding the correct chain length and node count answers requires including the $(n + 1)$ and $(n - 1)$ factors. If we omit them from the space chain length determination process, then a clear result will not be found.

6.5 Node solution stages

Appendix II, Theoretical Ratio of the Gravitational Force to the Electromagnetic Force between Two Electrons, provides details of the solution for node counts and chain segment lengths. Four solution stages exist where each stage provides an exact answer that the next stage requires. The result which I call the 'ggee ratio' indirectly produces the ratio calculated from 2022 CoData values. The result falls about 1.85 sigma above the calculated value. The driving constant for sigma is the gravitational constant which likely has its sigma understated. The only constant inputted to this process is *alpha*. *Alpha* has about twice as many significant digits as Big *G*. This means that the 1969 value for *alpha* produces the 2022 accepted value for Big *G* with full precision.

The stages needed to find node counts and chain segment links, and to perform the ggee ratio calculation are:

- Stage One – Determine the general structure of space and of the electron. This includes recognizing entanglement and its $n-1, n+1$ node count implications; and the end connecting link phasing of $0°$ or $180°$.
- Stage Two – Compute the node count and chain segment length for the electron segment. This calculation relies heavily on the irrational angle needed when a chain segment is twisted $90°$. It is important to understand the effect of the twist angle of the electron chain links. With reasonable stability measures, the solution for this stage is two orders of magnitude more stable than the second best solution. A precise single answer emerged.
- Stage Three – Using the node counts from Stage Two for the legs of a right triangle, determine the best rational approximation of the irrational hypotenuse. This is not a perfect right triangle. Originally, I avoided this calculation because it seemed to be a guessing game. Twenty years later (2016) I made a brief effort to find the hypotenuse. To my complete surprise a clear answer emerged on the first try.
- Stage Four – Calculation of the ggee ratio(a pure number with no units) was a trivial exercise after Stage Three. The calculations in this stage rely on the theoretical time dilation caused by the mass of the electron even though no time or mass units are involved. It was necessesary to include $alpha^2$ in the final result. The fine structure constant is also called the electron coupling constant. That is our usage.

Part III

Baryonic Matter

Chapter 7

Baryons

Part III may be skipped without loss of continuity. For DCT to be a useful model of the foundational structure it must be capable of structurally representing baryons. The structure needs to yield quantum numbers and an understanding of particle decay paths.

The structures shown for baryonic matter and decay paths are considered illustrative, rather than definitive. It is important that baryonic matter can be reasonably represented in DCT. For proof of concept, we consider particles in the major baryon octet.

7.1 Triangles

What is a triangle and why do we use them? You may recall that chain segment connecting ends align or counter align their phases. This provides a 'common denominator' for connections. Non-aligned phases in the connecting ends make building a structure much more challenging. It would be like building a physical structure with random sized corner angles.

Other reasons for using triangles include phase symmetry and the need for three connections at vertices. Three connections result from the need for connections to change from one chain link to another. A single 'side' connection can move as a single event. Multiple 'side' connections cannot. Take note that the 120° phase difference between connections does not conflict with the 180° angle sums for triangles. No geometry exists at the foundational reality level. This is not a geometric triangle.

How do triangles connect with the structure of space? They connect with chain segments at each vertex of the triangle. All of the connections connect to the same chain segment of space which facilitates a dynamic interaction. Recall that geometry does not exist. All vertices attach to the same space chain segment path.

A triangle seems to be the simplest structure that could exist in DCT, and the approach worked well. We call the triangle without the attaching chain segments the *core triangle*.

7.2 Baryon quantum numbers

The baryon quantum numbers are:

- Strangeness – *Strangeness* comes directly from the sum of the donut links contained in the core triangle. Strangeness equals the sum of the donut links minus 17.

- Isospin – *Isospin* comes from the baryon *core triangle* and the segments connecting the vertices to space chain segments.

  The core charge at each triangle vertex produces *core isospin*. *Core isospin* equals the core charge multiplied by 3/4.

  In addition to the core isospin, each of the connecting segments from a triangle vertex to space contributes isospin equal to its charge.

- Baryon spin – *Baryon spin* comes from the isospin contribution of each core vertex and its companion attaching chain segment. There is one important feature. The attaching chain segment usually connects with a space chain link with matching phase. An exception to this happens with the Xi^0 particle.

  The decay of the Xi^0 baryon provides the clue needed to understand how spin comes from isospin. One of the Xi^0 vertices connects with a counter-phase chain link (i.e., two links separated from the in-phase chain link). The Xi^0 decay to $Lambda^0$ retains the isospin at each vertex, but electromagnetically moves the counter-phase connection to an in-phase connection. This changes the spin by twice the isospin of that connection. The Xi^0 decay is the only major baryon decay that happens electromagnetically.

  It is possible that other connecting legs also attach counter-phase, but are hidden because their isospin is zero; or, cancels the isospin in another counter-phase attachment.

- Charge – *Charge* equals minus the number of *short-side* twists in a chain segment. Recall that the apices are either $0°$ or $180°$ in order to form a consistent connection with an external space chain segment. Charge can be either negative (short-side) or positive (long-side). In some cases a double charge (double twist) occurs.

- Hypercharge – *Hypercharge*, Y, for the baryon octet equals B plus S. This equals the net charge in the core triangle. Note that the total core charge remains constant for all three states.

The baryon quantum numbers are (continued):

- Color – *Color* corresponds with what I call *state*. Suvankar Majumder of Kolkata, India pointed this out to me.
- Quark – A *Quark* is the combined average effect of two sides of the core triangle and the connecting leg attached to the same vertex. The core charge for a quark equals the average over three states of the sum of the charges on the two sides. The external charge equals the charge on the connecting leg. For a strange quark the core charge is $-4/3$ and the connecting leg charge is $+1$. For an up quark the core charge is $+2/3$ and connecting leg charge is 0. For a down quark the core charge is $+2/3$ and connecting leg charge is -1.

7.3 Three states

Three states exist in the core triangle due to the external vertex connections moving to the next link in the core perimeter chain segments. This means that *spin* must be present in the core triangle. The total net charge in all three sides, called *hypercharge*, remains constant. *Three states* leads to fractional 1/3 charge units.

7.4 Core triangles

The number of space elements contained in a *core triangle's* perimeter vary directly with the Strangeness number. The number of space elements equals 17 plus the negative Strangeness number. Look at the core triangles in the next page.

The numbers shown equal the space element count in each core triangle side. The side charge can be 0, -1, +1, -2, or +2. Core triangles can have all vertices in the same phase, 0° or 180°. Alternatively, one of the vertices can be in the opposite phase as the other two vertices. If a side has two ends of a side in phase, then 4 elements would be zero charge. If a side has two ends in opposite phase, then 6 elements would be zero charge. Differences from zero charge can be positive or negative depending on the twist direction of the side.

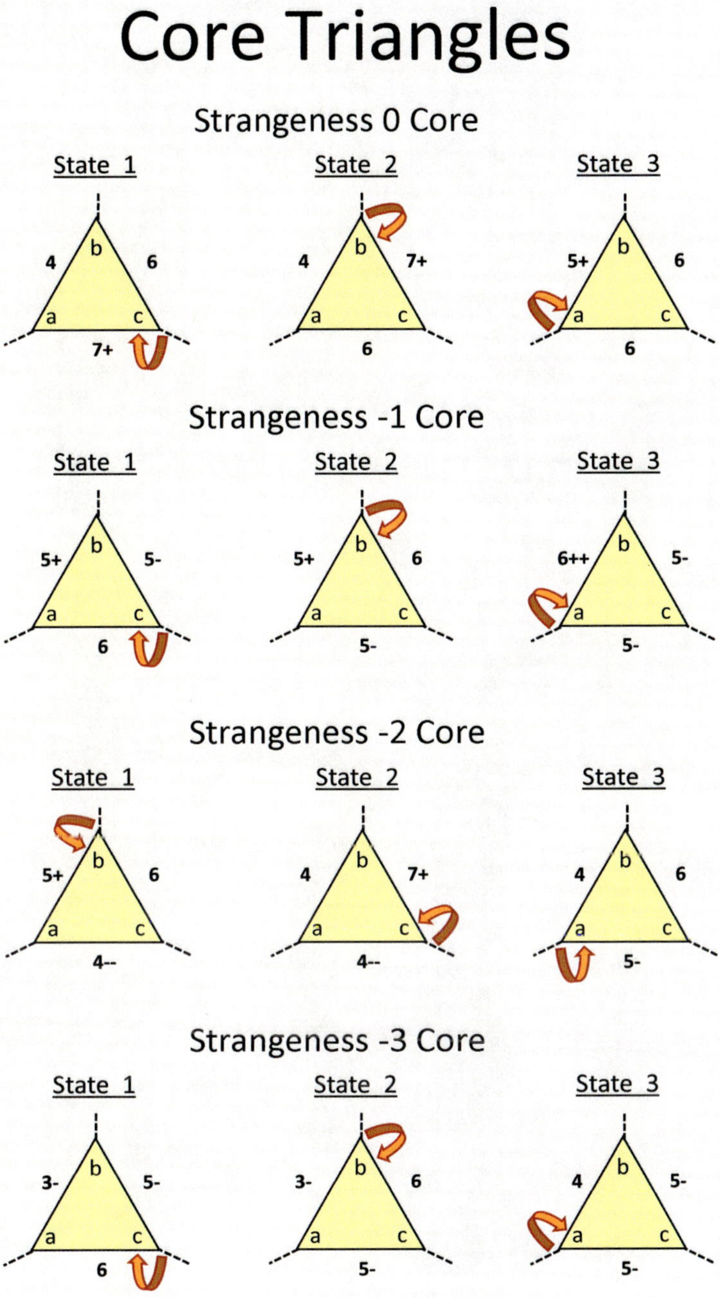

Figure 7.1: Strangeness Core Triangles.

7.5 Connecting chain segments

All apices of a *core triangle* attach to the same chain segment path in space. This needs to happen in order for the particle to travel a path through space. A *connecting chain segment* can have a charge twist of plus, minus, or neutral. The apex core isospin adds algebraically to the connecting chain segment isospin.

Usually the connecting chain segment attaches to a phase matching element in space. In the Sigma-0 particle we find that it connects counter phase. This apparently leads to an additive effect for the charge, but reverses the isospin effect on spin.

7.6 Gell-Mann–Nishijima formula

The Gell-Mann–Nishijima formula, originally given by Kazuhiko Nishijima and Tadao Nakano in 1953 and proposed independently by Murray Gell-Mann in 1956, states:

$$Q = I_3 + \frac{1}{2}(B + S) \quad \text{or,} \quad Q = I_3 + \frac{1}{2}Y \tag{1}$$

Separate the core and external to core I_3 components:

$$I_3 = {}_{\text{ext}}I_3 + {}_{\text{core}}I_3 \tag{2}$$

From section 7.2 substitute:

$${}_{\text{ext}}I_3 = {}_{\text{ext}}Q \quad \text{and,} \quad {}_{\text{core}}I_3 = \frac{3}{4} \cdot 2Y \tag{3}$$

Substitute equations (3) into (2) into (1):

$$Q = {}_{\text{ext}}Q + 2Y \tag{4}$$

This relationship bothers me somewhat because the charge balancing mesons leave tracks in the bubble chamber. Is it possible that the charged meson travels in the opposite direction due to the net charge impact of moving an external charge into the core?

7.7 Particle triangles

The following pages show the complete *particle triangles* for each particle in the primary baryon octet. Notice the Sigma-0 isospin sign reversal. Other vertices may also connect in a contra-spin manner, but have zero or canceling isospins.

Particle Triangles
(B = 1, S = 0, Y = +1, N = 17)

core state

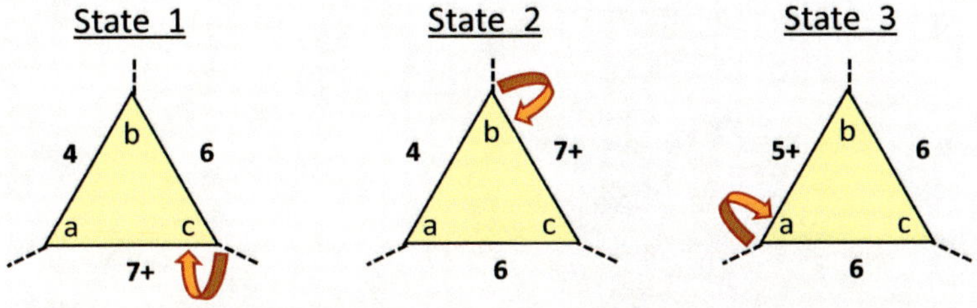

core side charge

state	ab	bc	ca
1	0	0	+1
2	0	+1	0
3	+1	0	0
average	+1/3	+1/3	+1/3

core apex sum

	a	b	c
charge	+2/3	+2/3	+2/3
I_3	+1/2	+1/2	+1/2

external apex charge (= I_3)

proton	0	-1	0
neutron	0	-1	-1

total charge, isospin & spin

proton-chg	+2/3(u)	-1/3(d)	+2/3(u)
$I_3 \rightarrow S$	+1/2	-1/2	+1/2
neutron-chg	+2/3(u)	-1/3(d)	-1/3(d)
$I_3 \rightarrow S$	+1/2	-1/2	-1/2

Figure 7.2: Strangeness 0 Core Triangle.

Particle Triangles

(B = 1, S = -1, Y = 0, N = 16)

core state

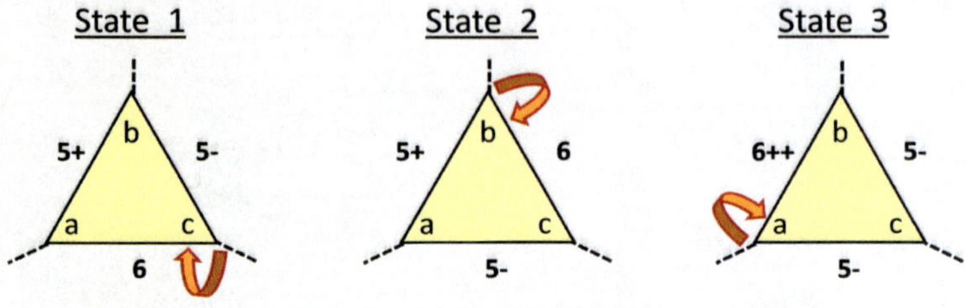

core side charge

state	ab	bc	ca
1	+1	-1	0
2	+1	0	-1
3	+2	-1	-1
average	+4/3	-2/3	-2/3

core apex sum

	a	b	c
charge	+2/3	+2/3	-4/3
I_3	+1/2	+1/2	-1

external apex charge (= I_3)

Λ^0	0	-1	+1

total charge, isospin & spin

Λ^0 charge	+2/3(u)	-1/3(d)	-1/3(s)
$I_3 \rightarrow S$	+1/2	-1/2	0

Figure 7.3: Strangeness -1a Core Triangle.

Particle Triangles

(B = 1, S = -1, Y = 0, N = 16)

core state

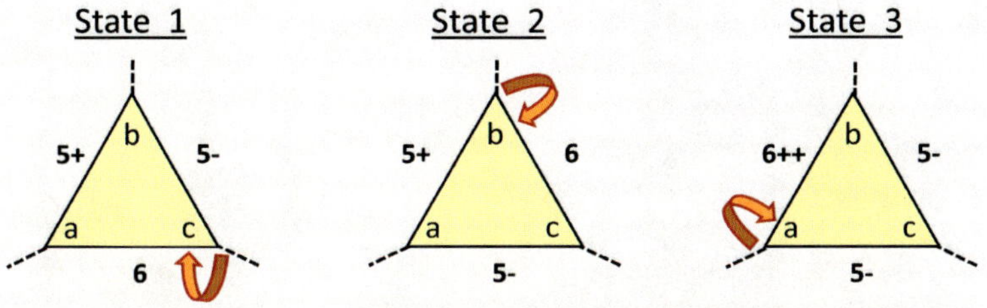

core side charge

state	ab	bc	ca
1	+1	-1	0
2	+1	0	-1
3	+2	-1	-1
average	+4/3	-2/3	-2/3

core apex sum

	a	b	c
charge	+2/3	+2/3	-4/3
I_3	+1/2	+1/2	-1

external apex charge (= I_3)

Σ^+	0	0	+1
Σ^0	0	-1	+1
Σ^-	-1	-1	+1

total charge, isospin & spin

Σ^+ charge	+2/3(u)	+2/3(u)	-1/3(s)
$I_3 \rightarrow S$	+1/2	+1/2	0
Σ^0 charge	+2/3(u)	-1/3(d)	-1/3(s)
$I_3 \rightarrow S$	+1/2	-1/2 → +1/2	0
Σ^- charge	-1/3(d)	-1/3(d)	-1/3(s)
$I_3 \rightarrow S$	-1/2	-1/2	0

Figure 7.4: Strangeness -1b Core Triangle.

Particle Triangles

(B = 1, S = -2, Y = -1, N = 15)

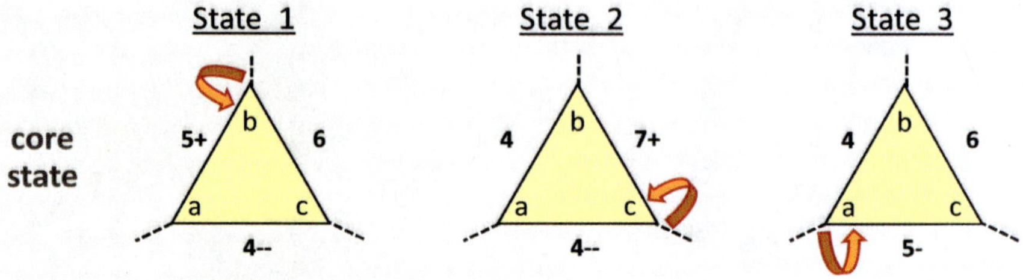

	state	ab	bc	ca
core side charge	1	+1	0	-2
	2	0	+1	-2
	3	0	0	-1
	average	+1/3	+1/3	-5/3

		a	b	c
core apex sum	charge	-4/3	+2/3	-4/3
	I_3	-1	+1/2	-1

external apex charge (= I_3)	Ξ^0	+1	0	+1
	Ξ^-	+1	-1	+1

total charge, isospin & spin	Ξ^0 charge	-1/3(s)	+2/3(u)	-1/3(s)
	$I_3 \rightarrow S$	0	+1/2	0
	Ξ^- charge	-1/3(s)	-1/3(d)	-1/3(s)
	$I_3 \rightarrow S$	0	-1/2	0

Figure 7.5: Strangeness -2 Core Triangle.

Particle Triangles

(B = 1, S = -3, Y = -2, N = 14)

core state

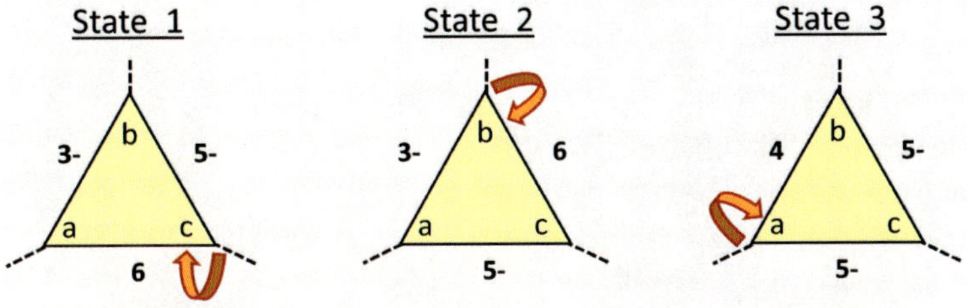

core side charge

state	ab	bc	ca
1	-1	-1	0
2	-1	0	-1
3	0	-1	-1
average	-2/3	-2/3	-2/3

core apex sum

	a	b	c
charge	-4/3	-4/3	-4/3
I_3	-1	-1	-1

external apex charge (= I_3)

Ω^-	+1	+1	+1

total charge, isospin & spin

Ω^- charge	-1/3(s)	-1/3(s)	-1/3(s)
$I_3 \rightarrow S$	0	0	0

Figure 7.6: Strangeness -3 Core Triangle.

7.8 Baryon decay paths

The illustrations on the following pages show the core triangles, external charges, and decay paths for the baryons in the major octet. All decays are considered to be started externally for particle collisions from a change in a connecting segment leg. Even the collision decays that result only in a core change are considered to be started externally from changes in charge potential. Another class of particle changes would occur from neutrinos incorporating themselves into the molecular structure. Figure 7.1. shows the core triangles for each strangeness number, S. Note that N, the number of chain links (space elements), decreases by 1 for each decrease in S. Apex phases are either 0° or 180°.

7.9 Baryon decay triangles

The *baryon decay triangles* provide a visual for many of the baryon decay paths. The concept of a triangle only provides a visual organization of the core relationships. There is no geometry at this level.

No significant attempt has been made to reconcile this to the Standard Model. Such efforts seem challenging. Exercise a degree of skepticism when looking at these. They are meant more for *proof of concept* than for an accurate and definitive explanation of the processes.

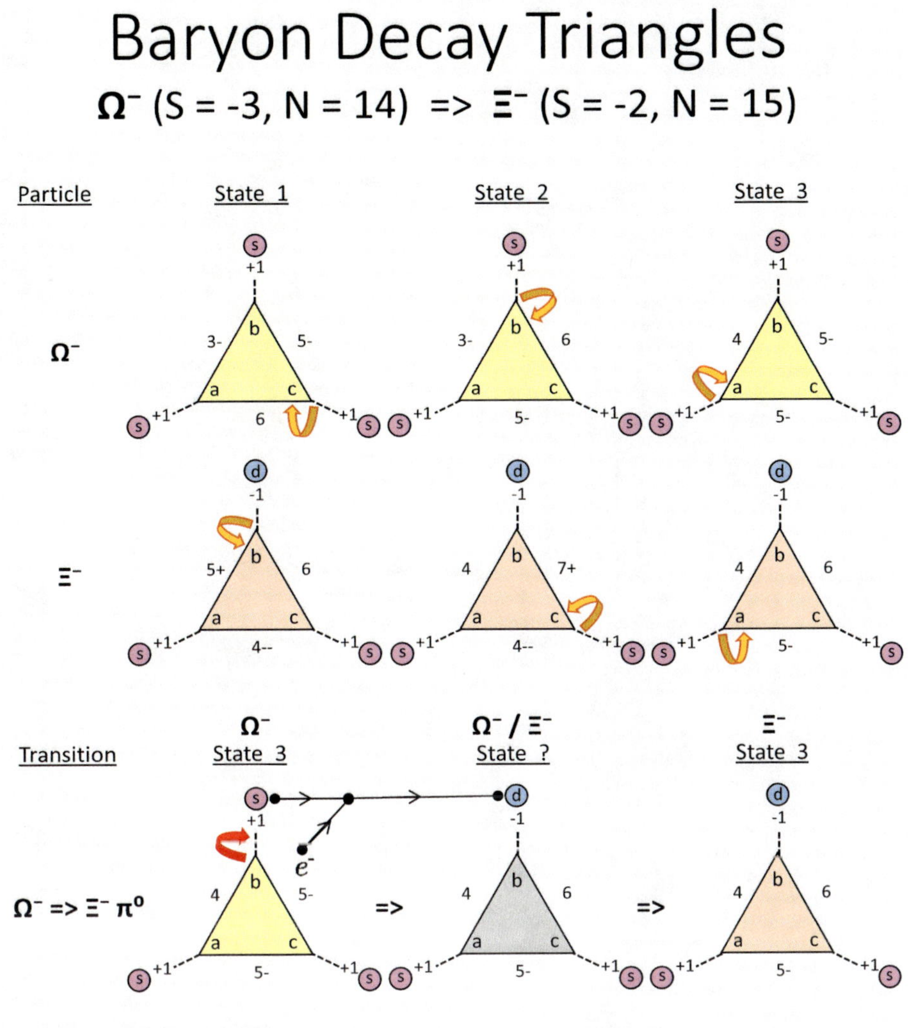

Figure 7.7: Omega-minus to Xi-minus decay.

The Omega-minus decays a strange quark into an up quark immediately balanced by adding an electron to change the up quark to a down quark. The strange quark decay moves a plus charge from a connecting leg into the core where it has a double effect. The electron added to the connecting leg balances the charge.

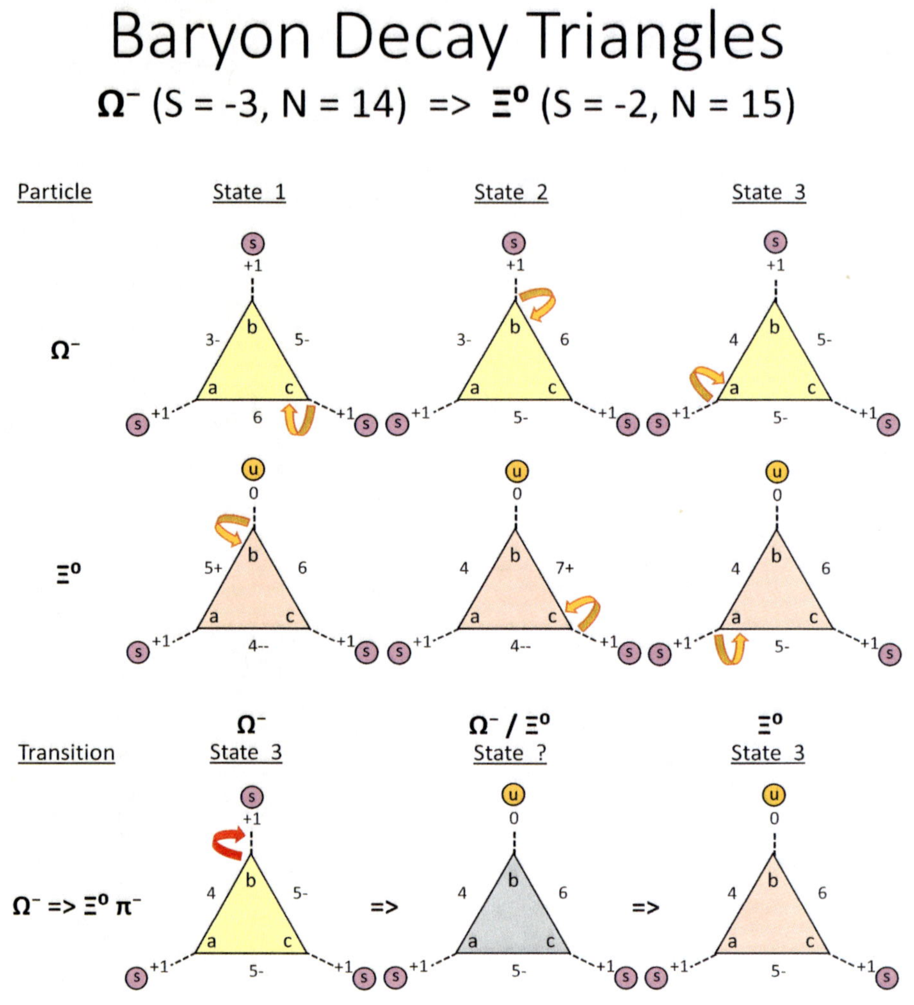

Figure 7.8: Omega-minus to Xi-zero decay.

The particle decays by simply moving the *ab* connecting side one space element farther out the connecting leg. The pi-minus would be attracted to the resulting more positively charge particle. The angle of approach for the pi-minus is not sufficient to allow it to penetrate into the particle and become a charge on the connecting leg.

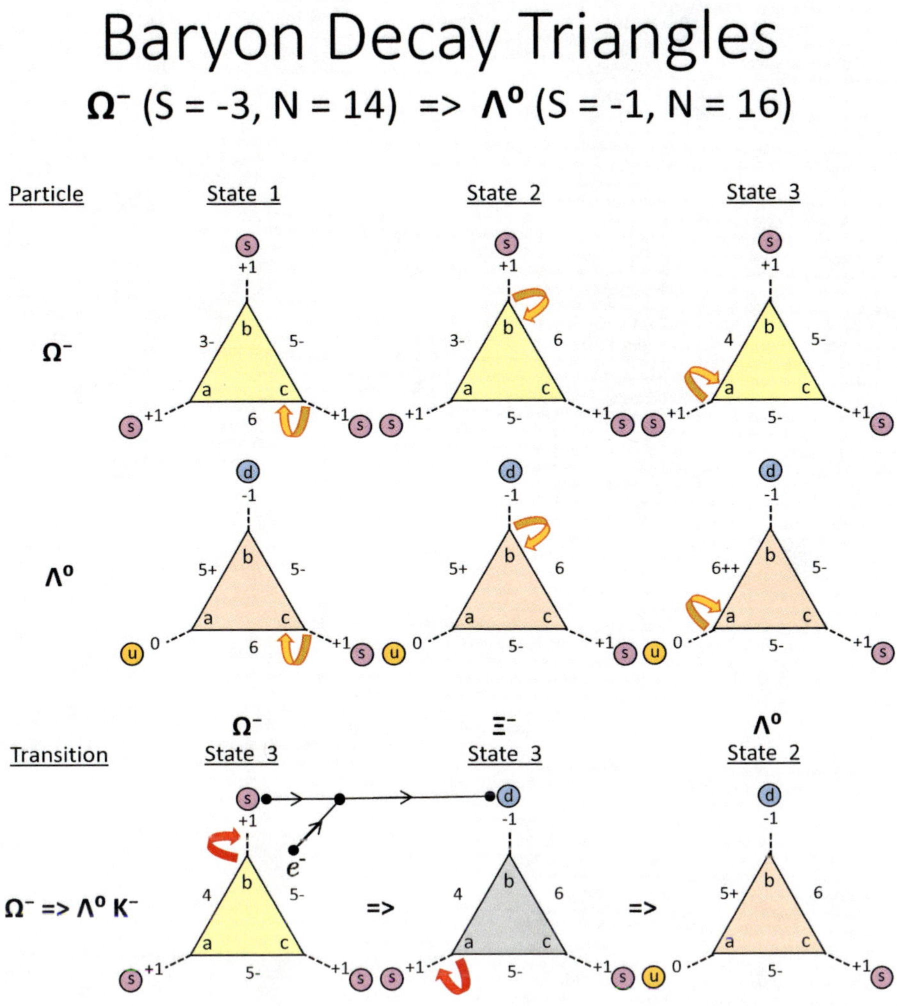

Figure 7.9: Omega-minus to Lambda-zero decay.

This is an interesting decay in that it takes a two-stage decay path with Xi-minus being an intermediate particle. Possibly, it doesn't add the electron until the final stage in which case the intermediate particle would be Xi-zero.

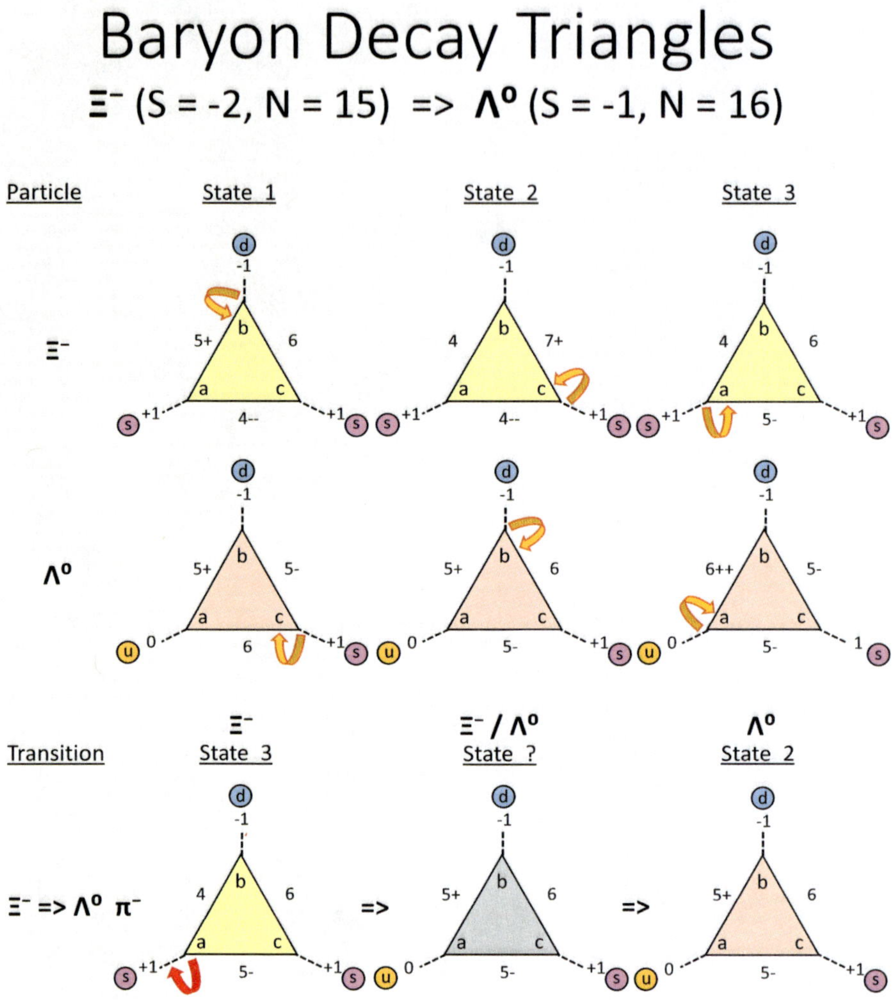

Figure 7.10: Xi-minus to Lambda-zero decay.

This is a straight forward decay of a strange quark into an up quark. The pi-minus needs to be an incoming negatively charged particle to balance the larger positive charge from moving the charge from a leg into the core.

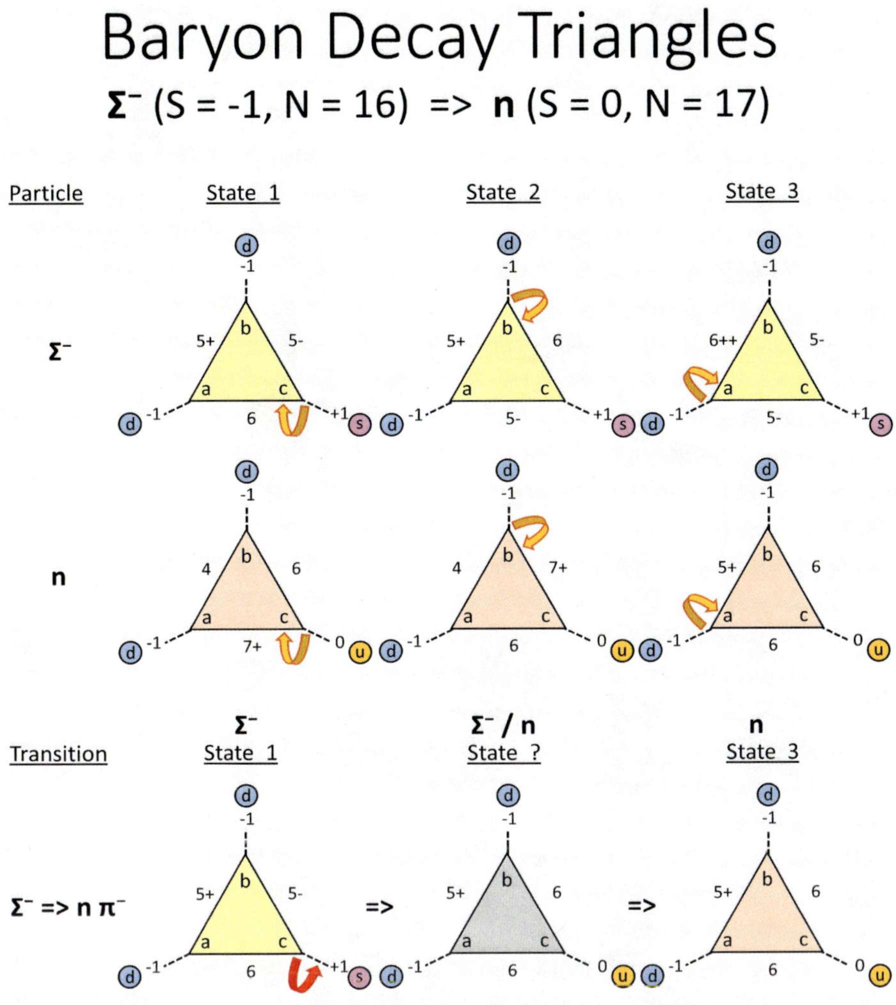

Figure 7.11: Sigma-minus to neutron decay.

This is another straight forward decay of a strange quark into an up quark. The pi-minus needs to be an incoming negatively charged particle to balance the larger positive charge from moving the charge from a leg into the core.

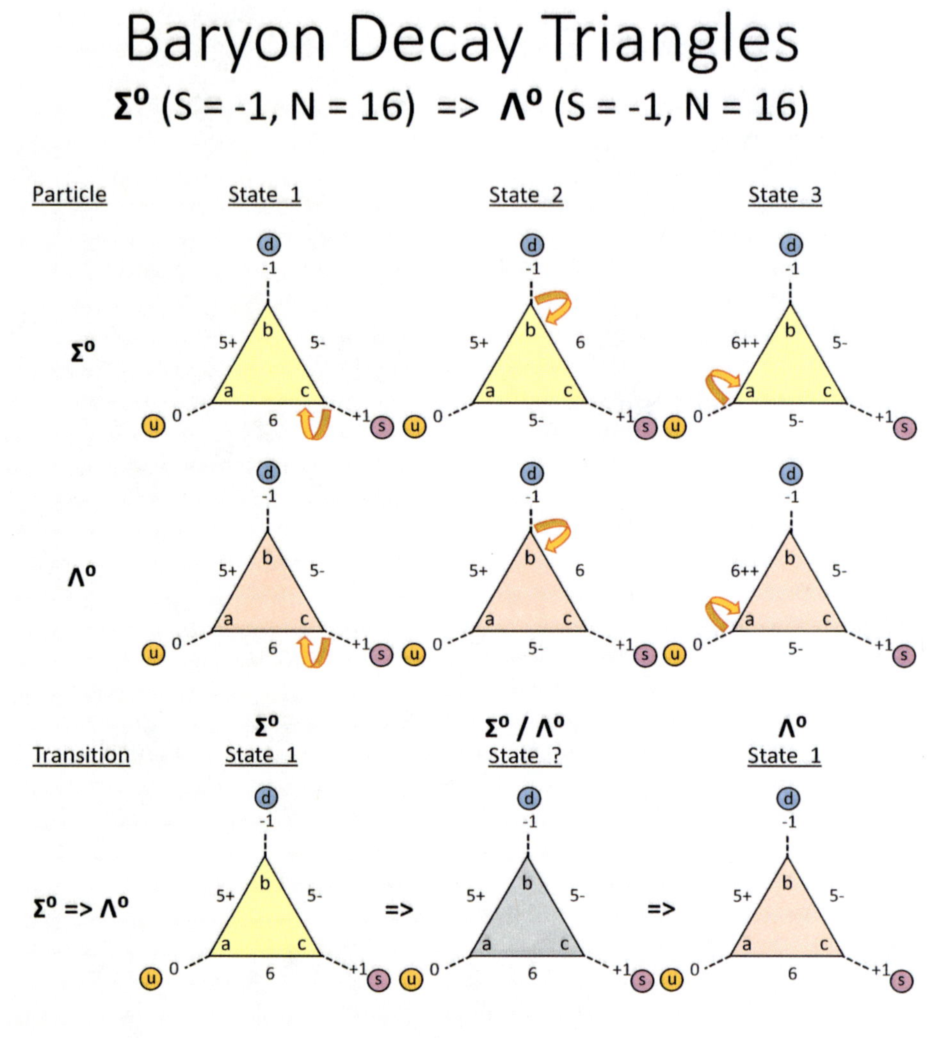

Figure 7.12: Sigma-zero to Lambda-zero decay.

This is the only major baryon that decays electromagnetically. There are no obvious changes except the spin changes from one to zero. Figure 7.12. helps us better understand the external segment connection of the core triangle to a strand of space. The Sigma-0 baryon has its apex b. connection with space 180° out of phase. The charge connects okay, but the spin becomes subtractive rather than additive. The electromagnetic nature of this decay provides clues about charge, spin, isospin and connection phasing.

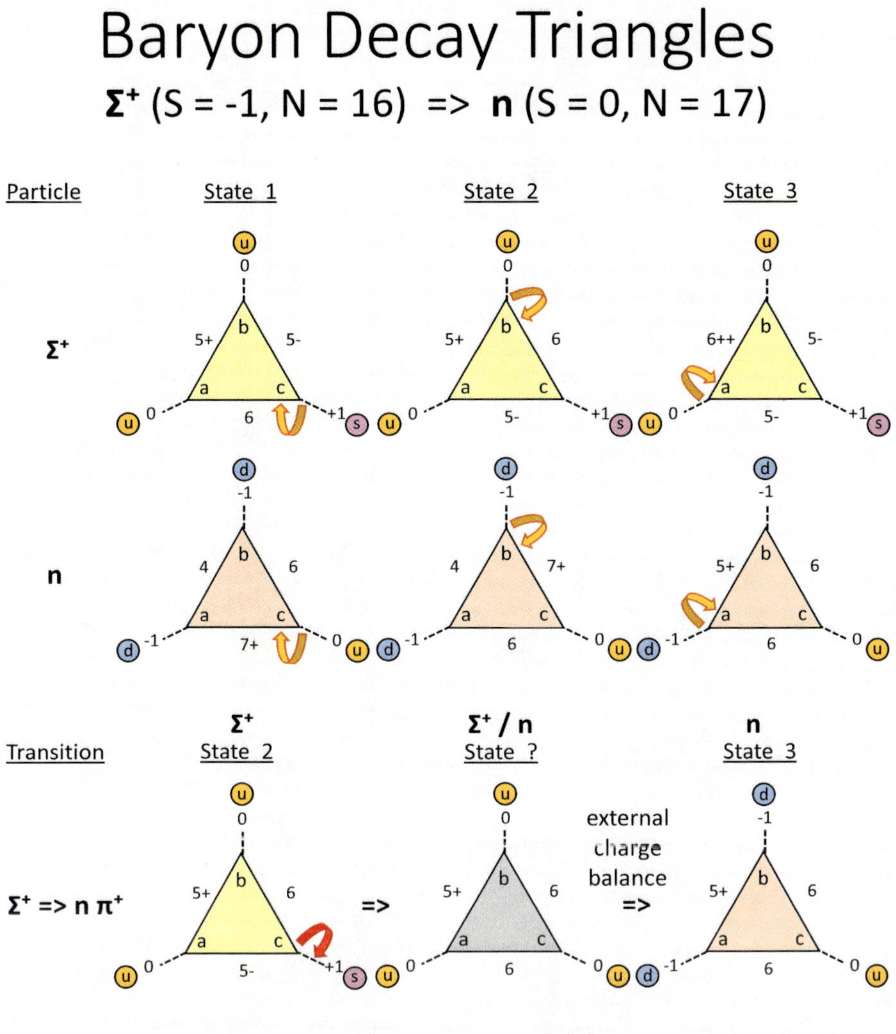

Figure 7.13: Sigma-plus to neutron decay.

This is a puzzling decay. I don't know how the two up quarks become down quarks. I label it 'external charge balance', but don't really know how the underlying process progresses. The Sigma-zero particle has its apex *b*. connection with space 180° out of phase. Possibly, something like that is happening with the Sigma-plus particle as well.

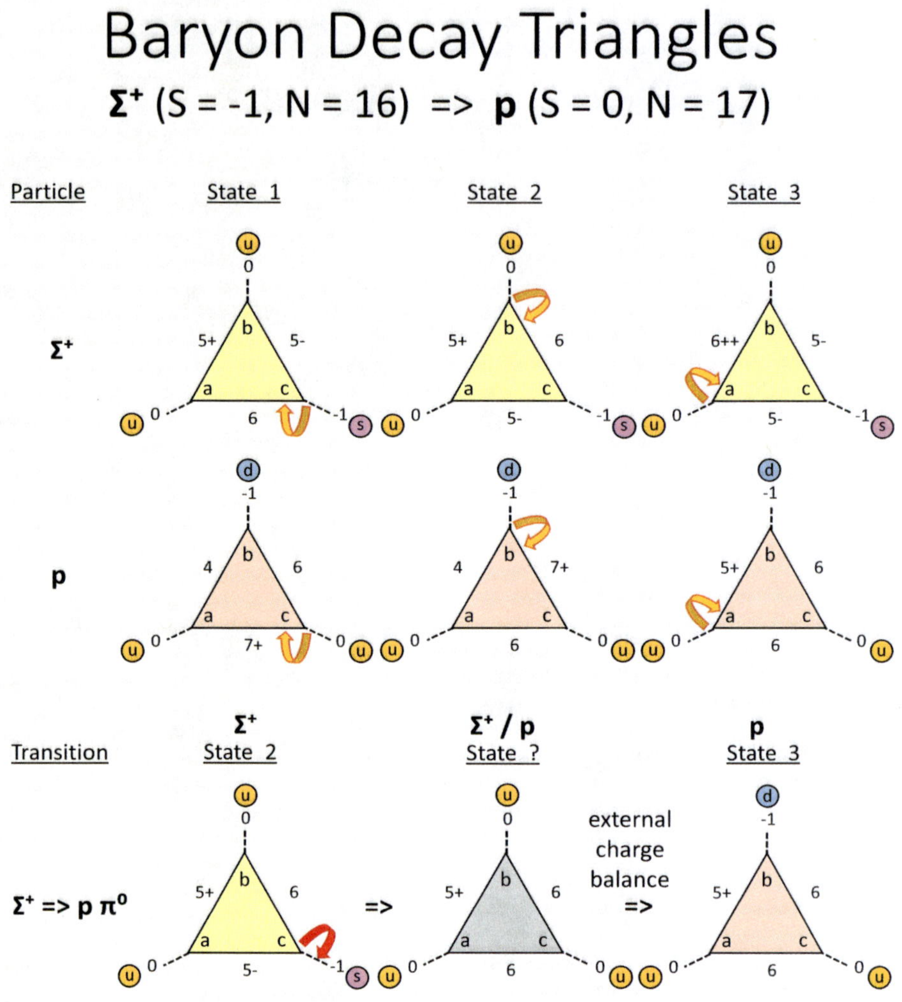

Figure 7.14: Sigma-plus to proton decay.

This decay is similar to the Sigma-plus to neutron decay; except, only one of the up quarks changes to a down quark. I label it 'external charge balance', but don't really know how the underlying process progresses. The Sigma-zero particle has its apex *b*. connection with space 180° out of phase. Possibly, something like that is happening with the Sigma-plus particle as well.

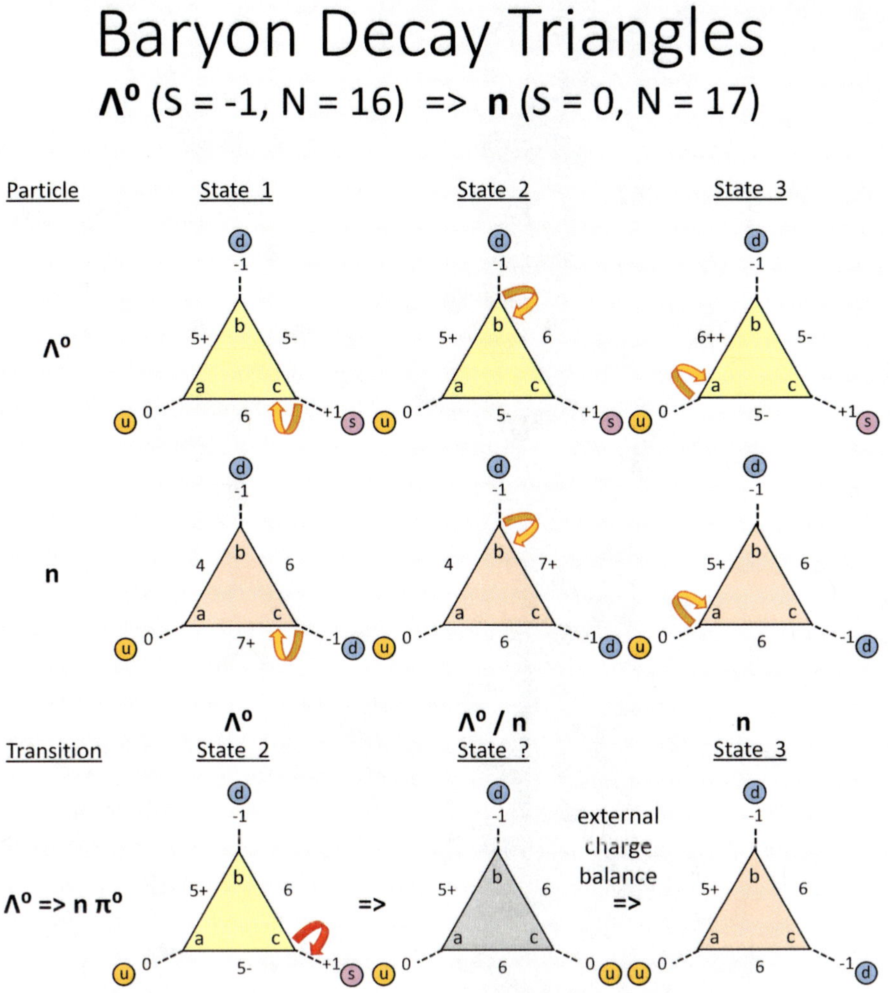

Figure 7.15: Lambda-zero to neutron decay

This decay is similar to the Sigma-plus to neutron decay; except, only one of the up quarks changes to a down quark. I label it 'external charge balance', but don't really know how the underlying process progresses. When the strange quark changes to an up quark, the effect of the positive charge doubles in the core from what it was in the connecting leg. This higher positive charge could attract an offsetting negatively charged electron with sufficient energy to penetrate the particle configuration.

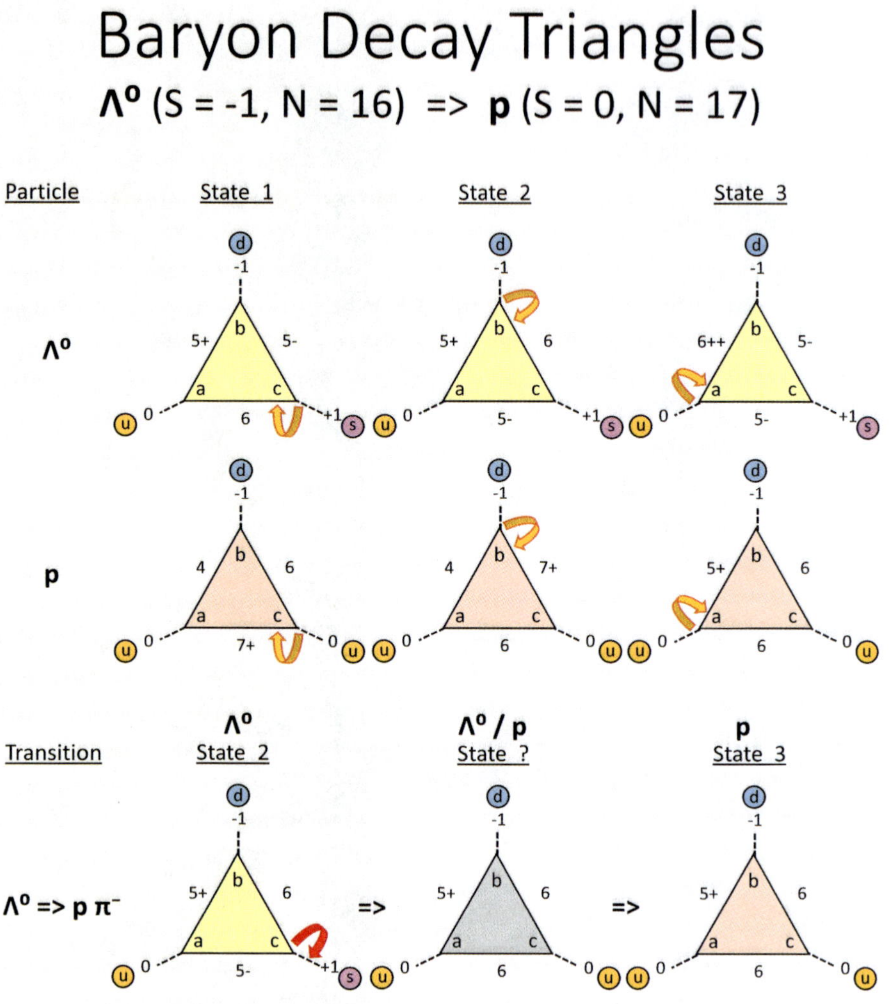

Figure 7.16: Lambda-zero to proton decay.

This decay is similar to the Lambda-zero to neutron decay; except, only the strange quark changes to an up quark., the effect of the positive charge doubles in the core from what it was in the connecting leg. This higher positive charge could attract an offsetting negatively charged electron but without sufficient energy to penetrate the particle configuration.

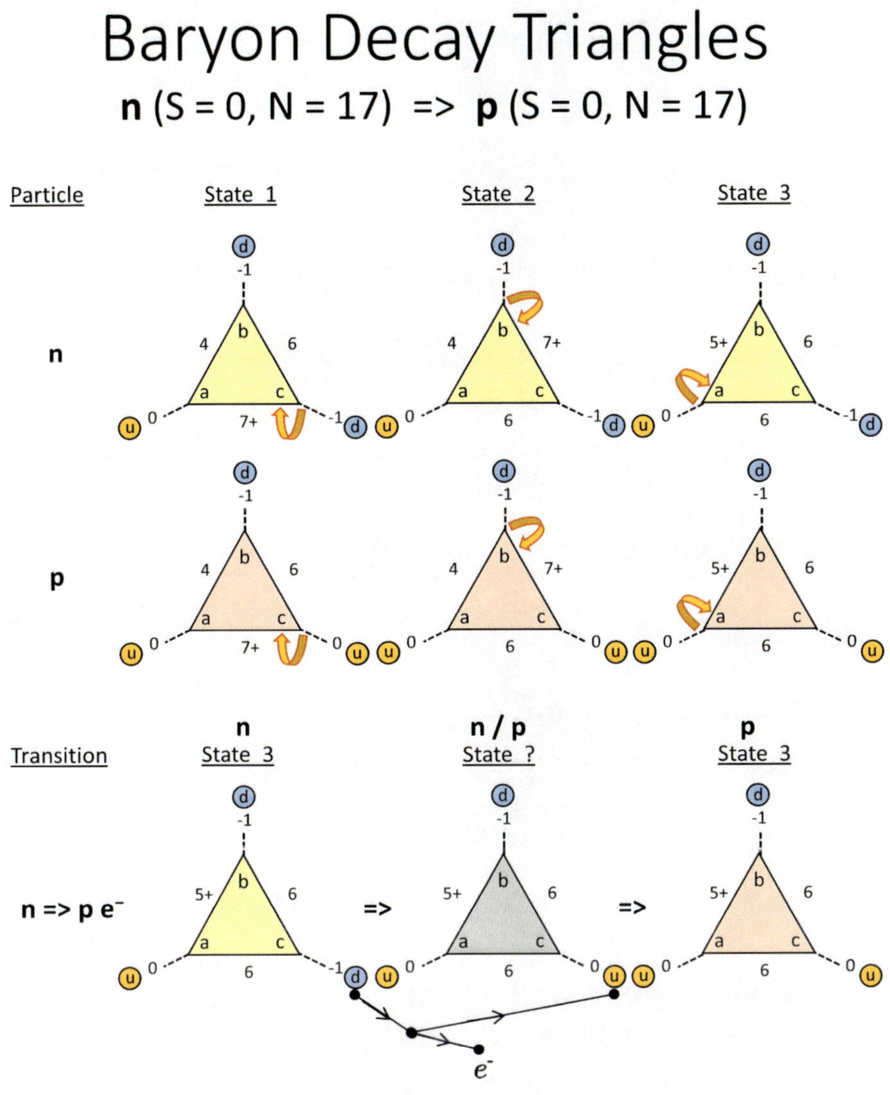

Figure 7.17: neutron to proton decay.

The beta decay shown simply expels an electron from a connecting leg of the neutron to change it into a proton. It does this by a space element penetrating into the particle configuration. This cancels the negative charge in a connecting leg and creates the external electron.

Part IV
Theory Implications

The implications part contains views that evolved directly from Donut Chain Theory (DCT). They are speculative, but indicated from the logic that led to DCT. I believe them to be largely correct.

Chapter 8
Quantum Physics

Interaction events between donut chain links must transfer incremental units of 'motion'. This needs to occur in multiples of discrete units in order for the chain links to remain synchronized. If the links lose synchronization then nothing is transferred because no interaction event occurs. This combined with the need for spin provides useful concepts for considering quantum physics.

8.1 Quantum packets

Packets *must* exist with DCT. If 'motion' were not transferred in integer multiples then the necessary node alignment (i.e., synchronization) will not occur. Without node synchronization nothing at all happens. An event only occurs when nodes on adjacent donut links occur in the exact same place at the exact same time. Position and time do not exist at the foundational level, but we use imaginary time and position in the donut chain link relationships. This seems to produce an exact phase relationship.

Note also that a quantum packet can only follow a single path. When a path branches, the packet will not distribute between the two branches. With larger packets consisting of multiple elementary packets, it is not clear whether certain conditions could cause some of the elementary packets to take one branch and the remaining elementary packet take the other branch.

When you think about donut chain links and related events try not to put too much of our physical reality into your view. For many years I thought of the imaginary motion of the donut that leads to nodes and phasing as actually happening in an imaginary sense. Now, I view it as an appropriate way to evaluate phasing between donut chain links.

8.2 Electron wave/particle duality

If particles are anything like little round balls then we have issues with the electron's wave/particle duality. There are no issues, only explanations that are needed. The electron chain segment has a twist due to its missing link. This twist causes a closed path of self-entanglement known as *spin* to also twist along its entire length.

At each space chain segment connecting link, the self-entanglement likely initially twists both connection legs. After one leg becomes a 'side' connection and the other leg the primary connection, an unwinding of the side connection occurs. It makes sense that the winding/unwinding of side connections would sometime synchronize with each other leading to something like a standing wave.

8.3 Uncertainty principle

Uncertainty seems unusual in our everyday reality. Surely, particles must have a position, even if we can't measure it. One can consider the connected chain segments that form the structure of space, where an electron consists of a chain segment with one fewer links. Where is the electron located? When it moves, it must move one chain segment at a time. There is no precise location or precise motion for the electron. *Uncertainty* arises naturally from the foundational structure. Our expectation of certainty creates the issue.

8.4 Double slit experiment

The *double slit experiment* with electrons lends itself to an interesting interpretation arising from the electron's need for self-entanglement. When the electron moves, a connected strand of chain segments must accompany it. With the double slit experiment this connected strand accompanies the electron through one slit, but the other portion of the connected strand returns via the second slit. If either slit is closed immediately after the electron passes through the first slit, then an interference pattern will not occur. In the case of a single electron, the position it strikes the background screen will be as if there were no second slit.

8.5 Quantum numbers

Quantum numbers (quantum features or states) for the primary baryon octet are discussed in Chapter III, Baryonic Matter. Mass calculations for the baryons have not been attempted. Chapter III presents a feasible way of explaining baryons in DCT. In order to consider DCT as a candidate for our foundational reality it needs to be able to represent the various quantum conditions.

Chapter 9
Fine Structure Constant

The reader may skip this chapter to avoid calculations.

Donut Chain Theory (DCT) provides a unique view of *why?* for the *fine structure constant* (a.k.a. *alpha*). The fine structure constant can be described as the probability amplitude that an electron emits/absorbs a photon, but much more remains to be said.

9.1 Fine structure constant

The *fine structure constant* is the most fundamental constant in Quantum Electrodynamics (QED). The Nobel prize was awarded to Richard Feynman, Julian Schwinger and Sin-Itiro Tomonoga for QED in 1965. *Alpha* is a pure number (i.e., dimensionless). It's known precision is twice that of the Newtonian Gravitational Constant, G (i.e., *alpha* has twice as many significant digits as G). This precision becomes important for the calculations in Appendix II.

The Fine Structure Constant also goes by other names. Originally, it was called Sommerfield's Constant. It is also called *alpha* and the *electron coupling constant*. DCT usage fits the *electron coupling constant* label.

9.2 Factoring the fine structure constant

I first looked at the fine structure constant *alpha* before understanding its importance to DCT. *Alpha* is readily factored into:

$$alpha = \frac{1}{137} \cdot \frac{3^4 \cdot 47}{2^5 \cdot 7 \cdot 17} = 7.297\ 353\ 2479 \times 10^{-3}$$

The 2022 CoData value for *alpha* is $7.297\ 352\ 5643(11) \times 10^{-3}$, where 11 is the standard error in the last two digits. Our error is 6836 in the last four digits. 620 times the standard error is huge.

Any decimal number can be estimated by a fraction. Even if the estimate for the decimal number falls near a standard error, the fraction should be given no special significance without a compelling reason. Hence, our factors have no special significance in this form.

Fast forward ten years. This happens after solving the chain segment length for space and for the electron. They are 138 links and 137 links, respectively.

Multiply both the numerator and the denominator by 5 in the earlier fraction and restate it:

$$alpha = \frac{1}{137} \cdot \frac{3^4 \cdot 47 \cdot 5}{2^5 \cdot 7 \cdot 17 \cdot 5} = \frac{1}{137} \cdot \frac{(138^2 - 3^2)}{(138^2 - 2^2)}$$

Our fraction now takes on a special significance. It still has the huge error in value, but lets us know that there could be more that needs to be found. When we examine the component factors in the fraction, they lead us to a better understanding.

9.3 The missing pieces

Fast forward another twenty years. This happens after solving for the *ggee ratio*. The solution required a factor equal to $alpha^2$. At the time I had forgotten that the fine structure constant is also called the electron coupling constant. That is exactly our usage.

We have a huge error in determining *alpha* from the earlier factors. How can this be addressed? Adding more factors without a compelling basis would be no answer at all. Instead we turn to *alpha*'s usage in the *ggee* formula. What aspects of the relationship have been omitted? Can those aspects account for the errors in the earlier expression? If so, getting an exact expression for *alpha* may be within reach; albeit, perhaps a difficult reach.

9.4 Components of alpha

Adding a propagation factor, $propfactor$, to *alpha* gives:

$$alpha = \frac{1}{137} \cdot \frac{(138^2 - 3^2)}{(138^2 - 2^2)} \cdot propfactor$$

We will individually consider each of the three *alpha* components. 137 in the first component deserves special mention. One might think that DCT's 137 chain links in the electron chain segment came from *alpha*. They did not. *Alpha* did not enter into DCT until the final calculation step for the *ggee ratio*. Conversely, it does seem that *alpha* comes from the electron's chain link count.

Examining the components of *alpha*:

- 137 – the attached chain segments of space can only interact with one end of the electron chain segment at any one instant. Assume this happens less frequently than the electron's internal chain link interactions. The electron's chain link motion would tend toward a uniform distribution. If it were perfectly uniformly distributed, then the spatial interaction would experience exactly 1/137 of the electron's internal motion. Instead, it experiences a slightly reduced amount.

- propagation factor – *propfactor* accounts for the nonuniform distribution of motion in the electron chain segment. This slightly reduces the 1/137 component of *alpha*. This calculation is nontrivial. Perhaps an especially skilled mathematician could accomplish it. I would suggest starting by assuming that a stationary electron's interaction with space happens alternately on each end at regular intervals. Hopefully, *propfactor* would suitably correct the huge error in determining *alpha*.

- main factor – the main component factor only indirectly affects *propfactor* by affecting the frequency of interaction with space. The numerator occurs in the space chain segments. They have an even link count with no twist. This allows alternating links to support 141 or 135 nodes and still maintain exact alignment with each other (i.e. a contact angle of zero). Note that both ends of the space segments would have the same node count. Note that the numerator contains 3^4. This helps synchronize chain segment end links which have 3 segments attached to the same link. It is conceivable that there are additional 3's in both the numerator and denominator that cancel. A greater number of 3's helps maintain event propagation across a greater number of chain segments.

  The denominator associated with the electron chain segment may also be associated with the self entanglement closed path. It could also benefit if canceling 3's existed in both the numerator and denominator. In this manner, all of the baryon triangle side counts would be present. If the strand to which the baryon connects is a part of an entanglement path, then factors would be present that matched each of the triangle's side counts. This helps synchronization.

Chapter 10
Gravitational Physics

Gravity accounts for much of our universe's behavior. General Relativity (GR) extends Newtonian gravity in mathematically beautiful ways. Many have tried to extend GR to the quantum level with limited success. Others have tried to extend it with dark matter to explain spiral galaxy rotation curves.

10.1 What causes gravity?

To quote Rod Serling('Twilight Zone' from long ago):

> You're traveling through another dimension, a dimension not only of sight and sound but of mind.

The concepts that explain gravity fall outside our normal thinking. They extend from thoughts that have come before, but require pausing your preconceived notions before listening to the explanation. Conventionally, matter leads to gravity which then leads to time dilation(clocks running slower). This order is wrong. Matter leads to time dilation which then leads to gravity. Many odd sounding concepts flow from this revised order.

10.2 Gravitational time dilation

Clocks run more slowly in stronger gravitational fields. This is called *gravitational time dilation.* Look at the reordered causal process. Let's say that matter by some mechanism slows the local clocks. The slower running clocks propagate their slowness outward. Under this view, gravity may be restated as the gradient of the rate of passage of time-squared. This view does more than simply restate accepted views.

If matter slows the passage of time in nearby space, what speeds the passage of time to maintain an equilibrium? The answer is that nothing speeds time back up. This simple concept can be difficult to grasp. What does it even mean? A person need not be concerned that slowing time will eventually stop. Time slows exponentially which means it continuously loses a portion of itself. The exponential series never ends.

10.3 Modified gravity

In 1983 Mordehai Milgrom presented a model of modified gravity called MOND (Modified Newtonian Dynamics). MOND modified gravity offers an empirical formula driven relationship to predict the rotation curves of spiral galaxies. Scientists have been pursuing the concept of *dark matter* to explain the flatness of the rotation curves.

The first time I read an overview of MOND gravity, I quite literally laughed out loud. The second time also evoked laughter. The third time I became convinced that modified gravity warranted a closer look. Two years and three tries later, I became sold on something resembling MOND. Part of my reasoning was that dark matter had far too many degrees of freedom. Let's put a little dark matter here and a little there. We will eventually find exactly the right amount and the right location to reinforce our belief in GRT's (General Relativity Theory) prediction of gravity.

MOND has issues in its arbitrary nature and predictions at really large distances, but it offers us a chance to consider alternative modified gravity approaches. Section 10.2 introduces the concept of time continually slowing in the universe. This slowing causes some form of modified gravity. It also makes mathematical modeling challenging since physical concepts rely on the a constant rate of flow of time. I had hoped to finish a modified gravity paper to include in this book. Instead, a brief description will need to suffice.

The components leading to some form of modified gravity are:

- gradient of time dilation – the gravitational time dilation from the originating mass dissipates over the fabric of space. The gradient of time dilation produces gravitational acceleration.
- flywheel effect – the fabric of space over which the time dilation dissipates can be viewed as consisting of a network of flywheels. The rotation of the flywheel can be thought of as the passage of time. With this view it takes a certain volume of flywheels to counter the time dilation (drag) from the originating mass. As the time dilation propagates outward it loses a portion of its initial 'slowing'.
- gravity meets slower time – when the drag from gravity arrives at a distant location it has lost some of its effect due to the flywheel dissipation. This decreases gravity's effect. The gravity encounters a slower local time due to the passage of time while propagating. This increases gravity's local effect.

10.4 Quantized gravity

Section 10.3, Modified gravity, describes gravity in terms of continuous functions. Yet, DCT *requires* gravity to be discreet and quantized. What form do these discreet relationships take? The answer to this question will need to be a best guess.

It seems unlikely that the units and mechanisms for electromagnetism would accurately describe the units and mechanisms for gravity. The electromagnetic fields and gravitational fields behave quite differently.

The *ggee ratio* paper (Appendix II) mentions a possible slight bias in the ratio produced (see page II.xiv). This bias would be consistent with a small portion of the gravitational mass occurring in the self-entanglement closed loop that closely relates to spin. The *ggee ratio* would be lower if the spin loop contained part of the collision angle mismatch due to the squared impact of the angle.

When considering the thoughts mentioned above, it seems likely that many more nodes exist than the $(n+1) \cdot (n-1)$ mentioned in section 6.4. This does not affect earlier calculations, but does affect gravitational calculations.

More nodes mean smaller units of quantized gravity. The additional nodes seem to act independently from the electromagnetic nodes although a fractional relationship would exist between the frequency of synchronization. Gravitational synchronization would occur with much finer gradations. It seems possible that gravity could have wave-like oscillations. These waves would be different from electromagnetic standing waves that occur between some nodes in the closed loop spin/self-entanglement path. Gravitational waves would be part of the propagation of gravity rather than between nodes in the closed loop.

Chapter 11
Our Universe

DCT *requires* a universe that differs from popular views. The DCT model provides an alternative to the popular view.

11.1 Boundary growth

The DCT model originates from interaction between neighboring elements of a space structure. The exact synchronization of neighboring elements forms a central part of the model. The model does not possess a mechanism for inserting elements in the interior. All growth must occur at the boundary of a universe. Two universes cannot combine to form a larger universe except under circumstances that may not exist. The boundaries must be exactly synchronized in order to combine.

– expanding universe

An expanding universe was postulated to explain the red-shift of the light from distant stars. Expansion increases the separation between large bodies. The space between large bodies does not expand. Molecular level expansion does not occur. Large bodies do not expand.

– big bang and accelerating redshift

The red-shift from an expanding universe leads to the *Big Bang* view by extrapolating backward in time. The more recent discovery that the red-shift accelerates was explained with *dark energy*.

– decreasing entropy

The boundary growth universe exhibits decreasing entropy. The Second Law of Thermodynamics remains intact because the universe is an open system. The *Friedmann Equations* do not apply.

Growth at the boundary initially is not well-ordered. Anomalies in the boundary growth create energetic matter. Gravity from the boundary growth matter propagates inward from the boundary since there is no outward space. This initially doubles the gravitational effect of the boundary matter. The boundary matter accelerates toward the more inward matter which in turn accelerates outward. Later, the earlier boundary matter accelerates toward more outward matter that forms as the boundary grows.

– cosmic microwave background radiation

Cosmic Microwave Background Radiation (CMBR) traditionally results from a reflection of the Big Bang. With boundary growth the CMBR results from the highly energetic matter added at the boundary. Why would this happen? When elements of space join the universe, their 'motion' has not yet fully synchronized with the universe that they are joining. This leads to connections not conforming to the 'motion' of the space of the universe. These mismatched connections form energetic matter.

– multiverses

The process by which boundary growth occurs also forms the universe initially by combining unattached elements of space. This means there are many more universes, perhaps infinitely more. Each universe must be either right-handed or left-handed due to connectivity constraints. Every universe slows at the same rate due to gravitational time dilation. The gravitational constant develops from the 'geometry' of the electron identically in every universe.

Multiverses in DCT differ from the concepts that add a dimension for a parallel universe; or, from the concept of a Many Worlds quantum interpretation. The separate universes act as completely separate entities that do not affect each other except as explained in the Gamma-ray Bursts (GRBs) subsection below.

One should not worry about two universes annihilating each other. This can't happen. For many decades I thought that separate universes never interacted due to the lack of synchronization. Now, I think that regions with gravitational time dilation offer the possibility of two universes interacting, but not in a way that facilitates communication or space travel.

– gamma-ray bursts

Appendix III discusses the features of gamma-ray bursts which indicate they may connect separate universes. The connection occurs in bursts because a portion of the matter slows the faster universe locally, allows the synchronization of time, then moves into the slower universe. GRBs often are accompanied by binary neutron stars or black holes, both of which exhibit strong gravitational time dilation. Other GRBs may have lacked detection of locally strong gravity; or, they may connect with universes that have only slightly slower clock rates. Another possibility would be that our universe is the slower universe being connected by a faster universe.

An open question would be: "How do universes maintain their density of matter and energy with a GRB connection?"

11.2 Electron, choreographer of the universe

To understand how and why the universe formed as it did we look for what choreographed its formation. We conclude that all universes form in the same manner with the same fundamental constants and physical attributes. Other than handedness universes all grow in a similar manner. An exception to this could be the foundational structure connections that lead to three spatial dimensions.

– electron, the master choreographer

The electron acts as the master choreographer in all universes. Other sections of this book describe the reasons in detail. Section 4.3 examines the need for an electron. Section 5.3 explains why entanglement and spin *must* accompany the electron. Finally, page II.x in Appendix II details calculation of the electron's structure.

No other real choice exists for a universe's formation than having the electron at its root. In DCT, the foundational connections *must* produce three spatial dimensions. This must be so because our physical reality fits with a view of three spatial dimensions. I have found no reasoning for three spatial dimensions to occur. Since they do occur in our universe, they must be supported by consistent foundational connections. Does this mean a different universe would also contain three spatial dimensions? or not?

– electron and gravity

How could the meager electron control the gravitational constant? It seems that heavier matter would be the better candidate.

The electron's significance comes from synchronization needs. The electron itself has one fewer elements than normal space. This leads to a slight phase mismatch. An electron can achieve synchronization if time slows at a particular rate. Without synchronization, nothing at all happens. In this manner, the slowing of time leads to synchronization. The particular rate at which time slows results from the mass-energy density of the universe. The universe does not form without the particular slowing rate dictated by the electron.

– communicating between universes

Two universes both have electromagnetism, but they are not connected electromagnetically. Communication between closely synchronized universes with the same handedness could occur with neutrino signals if that were even possible. Figuring out how to communicate with neutrinos seems out of reach.

11.3 Three dimensions

In our everyday physical reality we take three dimensional geometry for granted. How can we get a three dimensional reality from the foundational structure? The answer to this question will seem absurd to most readers because we are accustomed to our reality.

The foundational structure has no geometry. It consists of connections and interactions between neighboring elements of space. Yet, we know that in order to be correct, it must lead to the three space dimensions of our normal reality. It must also lead to quantum physics and produce the orbitals used in chemistry. These requirements provide clues about the structural connections in our hidden foundation.

– **geometry**

The foundational structure has no geometry at all. It only has neighboring elements of space. How could that produce three dimensions? This seems especially difficult when one needs to reproduce geometric relationships, such as the precise length of a diagonal that connects opposing corners in a cube-like shape.

– **points of reality**

With the exception of neutrinos all physically detectable activity in the foundational structure occurs at the connections between chain segments. The intrasegment chain links remain hidden from physical observation. The physically detectable points could be considered *points of reality*. It is hard to imagine that points in foundational space lacking geometry can support geometry in our physical reality.

11.4 Distorted geometry

A severely distorted geometry in physical reality seems to mean that such a concept does not work. Consider FitzGerald's 1889 letter to the editor (Section 2.1) in which he voiced an explanation of why the Michelson-Morley (M-M) experiment returned null results. FitzGerald pointed out that the physical apparatus could be held together electromagnetically and respond the same as the light beam being tested. This is what also happens in the distorted geometry produced by foundational points of reality. Any distortion from the underlying foundation affects everything the same. For example, a shorter distance would be measured using a shorter ruler and yield the same numeric measurement. Our predictable geometry comes from a distorted geometry of which we remain unaware.

Figure 11.1 shows a crude example of connections in the foundational structure. This example conforms to a three dimensional view. The foundational chain segments could form knots or be severely deformed depending on one's point of view.

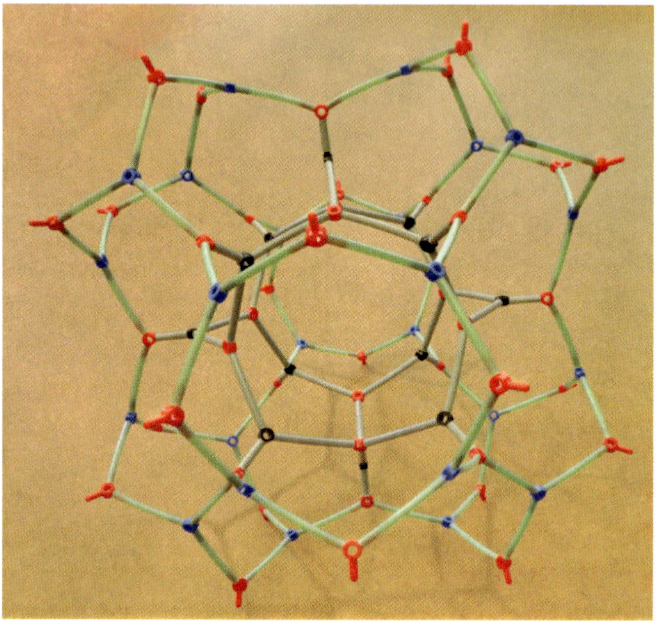

Figure 11.1: **Crude Example of 3-D Connections**

The example exhibits several noteworthy features:

- Chain Segments – The tubes represent *chain segments*. The colored three-way end connectors represent the end connecting links of the chain segments. All of the blue (or black) connectors represent 0° phase. The red connectors represent 180° phase. Particularly note that distortions in the length or direction of chain segments have no effect on the results.
- Phasing – The example contains paths that reconnect after eight chain segments. Chain segment paths need to reconnect in phase. The segment count needs to be a multiple of two.
- Three Dimensions – The example forms a crude three dimensional symmetric structure. The foundation needs to support three dimensions, but does not need to support other aspects of our geometry.

Clues to help us to understand details of the foundational structure may lie in the energy levels and pairings of electron orbitals. There are many visual representations of orbitals.

Chapter 12
Donut Chain Theory Validity

12.1 Problems

The donut chain segment calculations rely on an exact synchronization and an exact cancellation of opposing motion. No explanation presents itself to explain why this exactness happens. Such an exact requirement seems unnatural. I would expect some type of process that seeks an equilibrium.

A problem with DCT arises from very substantial departures from current views and beliefs. DCT offers alternative explanations for many phenomena. This may give the impression that DCT conflicts with reality. I don't believe it does. A correct understanding of the foundational structure must *never* conflict with reality.

12.2 Merits

DCT started as conjecture. All such theories start as conjecture. When a theory coincides with reality and seems to have no conflicts with reality it may become plausible or useful. Rarely does a theory come with a label of authenticity. It seems challenging to even imagine what such a label would be. DCT comes with such a label. The *ggee ratio* calculation provides the ratio reasonably within currently best available numbers. The calculation consists of several stages each relying on results from the previous stage. The *ggee ratio* that emerges provides a compelling reason to consider DCT as a correct foundational understanding.

Early in the process DCT *requires* self-entanglement and spin. It leads to a meaningful representation of baryonic matter. All of this seems too good to be true. That impression creates a hurdle.

12.3 Acceptance

DCT has no recognized acceptance. I doubt that it has ever been considered by anyone in a position to foster such an acceptance. I have no expectation that DCT will ever achieve acceptance. This is far from a tragedy. DCT could provide a tool to help physicists and philosophers, but it is not a be-all or end-all necessity.

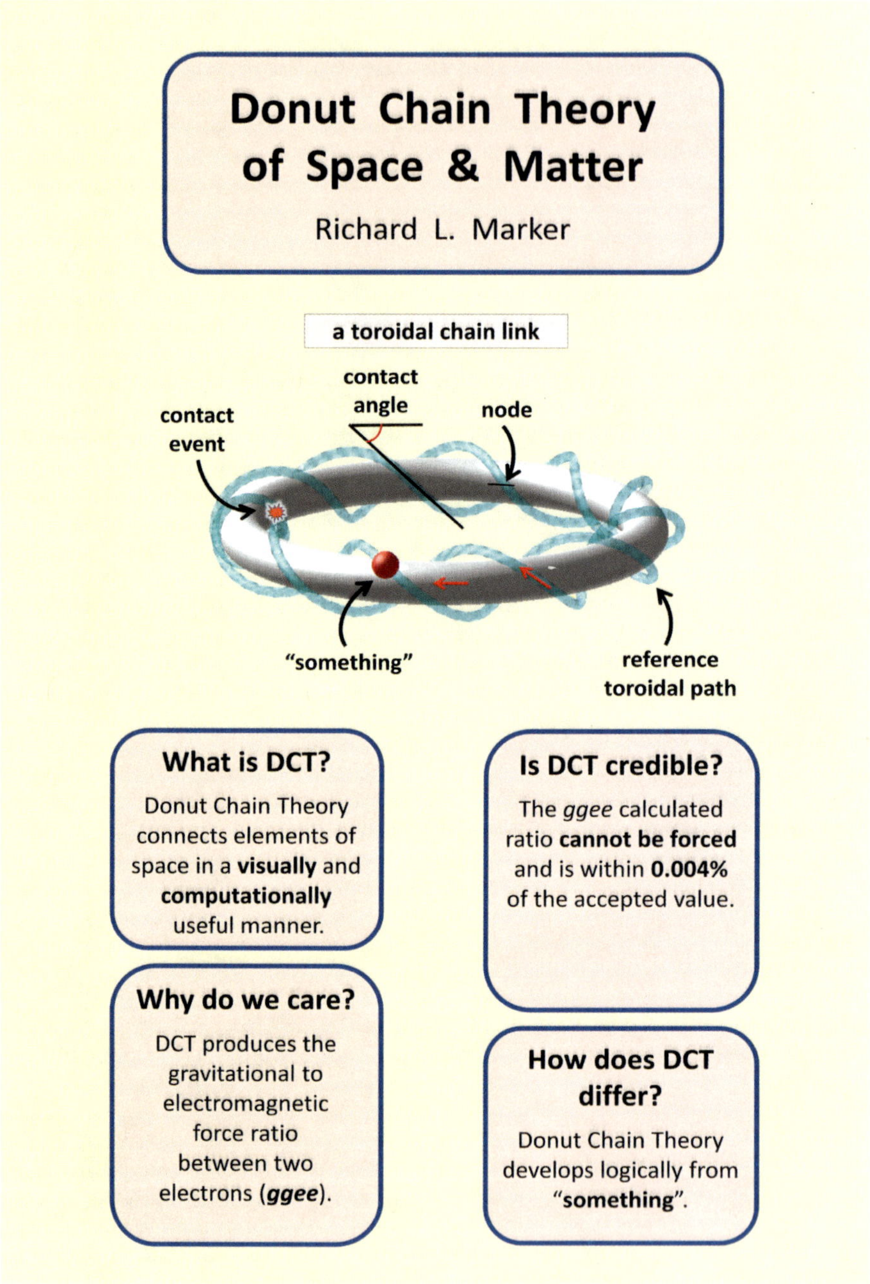

Figure 12.1: Donut Chain Theory

Figure 12.1. shows a poster overview of DCT. The details that support these claims are both complex and contradict some currently accepted views.

Part V
Theory Speculations

The speculations part contains views that did not evolve directly from Donut Chain Theory (DCT). The views have mechanisms of support in DCT, but should be viewed with skepticism. Nothing is meant to challenge a person's religious beliefs. These speculations form a compatible *back story* that helps complete my understanding.

Chapter 13
Minds

Our minds influence or control most of what we do in life. We may have barely scratched the surface of understanding the mind's limits or reaches. What does this have to do with foundational reality? Possibly a great deal; or, possibly nothing.

Most of the topics in this chapter present concepts based on anecdotal evidence. This introduces a great deal of subjectivity and room for other opinions. My views base from experiences, from consistency, and from the possible reaches of entanglement. Foundational reality seems to relate to this from both the anecdotal evidence of experiences and the mechanism of entanglement.

13.1 Sleep thinking

Within the past year I did an internet search on *sleep thinking*. To my surprise it is a recognized process and some approaches closely resemble my own approach. Generally, the effectiveness of sleep thinking is attributed to the mind working extremely efficiently during sleep. That is not my view.

My approach worked for seemingly unsolvable paradoxes. I would carefully identify the paradox while drifting off to sleep. Some mornings I would have an apparent answer that needed to be remembered immediately; or, be lost as fast as it was found.

The anecdotal results from *sleep thinking* made a believer out of me. I used these methods for seemingly unsolvable paradoxes in DCT. Every single crux thought came from sleep thinking. Only once did sleep thinking provide a wrong answer. That one time the answer seemed so beautiful that I felt endorphins. Two years later thanks again to sleep thinking I had to go back to the last major move and completely discard the beautiful thoughts. This was really difficult. Now, I can not even remember the false thoughts.

I like to think that the answers provided by sleep thinking came from others, perhaps even others from earlier times. This sounds more than a little odd, but it is how I like to believe.

13.2 Entangled minds

In order for the electron to move, DCT *requires* that it has a companion closed loop path in space. DCT calls this closed loop *self-entanglement.* It is directly related to spin. Some people believe the mind operates on quantum entanglement. If the closed entanglement loop links two or more particles it becomes quantum entanglement. This process could occur within the mind; or, between different minds.

Different minds could be entangled with each other in varying degrees. This is where it gets interesting!

13.3 Mind of God

The *Mind of God* title labels the process I consider to be consistent with DCT. It is not meant to offend anyone. It is also not intended to supplant the reader's personal views. Please consider it to simply be my personal belief label.

Consider the entanglement of different minds in varying degrees for various beliefs. Extend that thought to all of humanity. The many entangled thoughts and beliefs resemble a brain with connected neurons. I call this concept the *Mind of God.*

Since the individual minds contributing to the *Mind of God* have a memory, the overall mind could also have a memory. There could be a record of thoughts of unknown magnitude from those who have lived before.

Could the *Mind of God* include entanglement with the minds of other animals? It seems possible, but not likely to be of the same degree as that with other humans. Still, the degree of individual entanglements varies. It even seems possible that some people could be closely entangled with some animals.

Chapter 14
Religion and Culture

14.1 Religions

Religions vary in ways that seem irreconcilable. Which religion is correct? or, incorrect? I believe all religions are correct. How can that be possible?

– entangled groups

Groups of people may be closely entangled in particular thoughts or beliefs. Their shared religious belief system entangles the minds of the followers of that faith. Their religion serves them in a correct manner for them. In this way all religions are correct for the group that chooses to practice them.

– son of God

There may be room in the entanglement view for support of a *super entangled* person. Let's call this person the son of God. It seems within the realm of possibility that someone could be so entangled with a group of followers that they could perform miracles. Many consider healing through prayer real. A *super entangled* person could conceivably do this. Perhaps they could also do other miracles.

– good or bad religions

Good or bad comes from the eyes of the beholder. From the eyes of the believer a religion forms part of their belief system. If a majority of people consider a religion bad, it may be a cult; or, it may simply differ from their own belief system.

14.2 Cultures

There are many similarities between religions and cultures. Cultural norms often have some utility in the culture to which they apply. Cultural norms do not have to come from entanglement. They could come from societal behaviors. Distinguishing the source may well not be possible. The part that relates to DCT comes from the entanglement of people's minds.

– knowing right from wrong

Are we born knowing right from wrong? Assume for a moment that most people know right from wrong. Possible sources of such knowledge are:

- hard wired at birth – This suggests that brains evolved to form a hard wired brain that distinguishes right from wrong.
- enculturation – It seems logical that cultural influences would shape one's senses of right and wrong. This occurs over time.
- cultural entanglement – This has similarities to enculturation, but would be present at birth. Some studies indicate that a knowledge of good behavior is present at birth. A brain's entanglement with others in the culture could explain this.

– Inca child sacrifices

It was considered a great honor in the ancient Inca Capacocha ceremony for a child to be selected as a sacrifice to a deity. Today, most would view such a practice as horrific. Does a child know right from wrong in the Inca case? It could if *cultural entanglement* occurred before birth.

14.3 Kindness

To me the greatest achievement in life for which a person can be remembered is kindness. I have noticed that some cultures and some groups practice kindness. It seems contagious.

One group I particularly remember is the Loop Quantum Gravity group. It is refreshing to see and affirms one's faith in the humanity of others. Thank you for your kindness!

Closer to home I continue to be impressed by Bill Gates. During the early years of Microsoft I was not pleased by the company's business approaches. Now, the work the Gates Foundation does to fight HIV in poor nations has completely reversed my image of him. Thank you, Bill Gates, for caring!

There are many, many more who keep caring. It is much of what makes life a pleasure.

Chapter 15
Human Existence

The topics in this chapter provide a glimpse of where the concept of entanglement could lead. The sections may be thought provoking, but remember that they are rife with speculation.

15.1 Evolution

Charles Darwin's 1859 treatise on natural selection, *The Origin of Species*, may colloquially be spoken of as survival of the fittest. What if it has little to do with the reproduction of longer lived members? Consider the details passed in one's DNA. It seems extraordinary to me that so much detail could possibly be contained in strands of DNA.

Let's posit an alternative. Assume that matching strands of DNA in different members of a group have a strong degree of entanglement. Do the features of other living members provide the living roadmap for growth for the newer members? Interestingly, this means that members of a group who do not reproduce may still have an effect on the evolution of the group. It also means that evolution could advance at a rapid pace.

This alternative view of DNA could be useful in understanding how to better control cancer; or, combat difficult viruses. The extension of these concepts could be far reaching. Perhaps instead of looking for a drug that rapidly kills a virus, we should look for a drug that slowly kills a virus. Let the susceptible virus linger to provide continued replication before subjecting it to an early demise.

15.2 Reincarnation

The memory of the *Mind of God* (Section 13.3) provides the framework for considering reincarnation. Maybe retained thoughts of a predecessor manifest in the mind of someone living.

This does not predict or explain reincarnation. It suggests that claiming reincarnation may be more rational than it seems. Perhaps the night-time visit from a deceased person has an origin in the *Mind of God*; instead of being an invention of our dreaming mind. Don't dismiss a thought too quickly as your mind's invention.

15.3 Free will

Is the future predetermined by the present? If everything obeys definite physical laws that might seem to be the case. Contrariwise, one may argue that the probabilistic nature of quantum physics allows room for an indefinite future. I think quantum physics is foundationally determinate and not probabilistically determined. Regardless, I believe in limited free will. How could that be logical?

If some aspect of the universe is nondeterministic then the predetermined test fails on some scale. Does that mean human behavior is not predetermined? Regardless of what that means, I don't believe in predetermination. I also do not believe in a completely free will. I believe we have cultural, religious or legal constraints that influence our choices. Many of these factors would seem to come from our mind's entanglement with the *Mind of God*.

Thus, we have limited free will. The limiting factors control our behavior in a very real, but not absolute, manner.

15.4 Soul

I view the soul as our mind's entanglement both within itself and with the *Mind of God*. The origin of the entanglement seems to lie in the inherited genome. The ultimate destiny of the soul would seem to be the traces it leaves in the *Mind of God* (i.e. in the 'memory' of the living people that constitute the *Mind of God*). This memory would not need to be active any more than our own memories have everything readily available to our thinking process.

As described, our soul would be the parts of our entanglement that physically reside in our mind. Perhaps it could include some parts that do not physically reside in our mind. There could even be 'cloud storage' of the entanglement residing in our mind. In this case, the soul would not disappear on death. These are just speculations. Other people may have their own views that differ.

15.5 Intuition

The entanglement that produces the *Mind of God* could be viewed as a huge library of thoughts rather than a library of words. Many thoughts would be incomplete or contradictory. Connecting with this all-knowing thought library could produce *intuition*.

Instead of looking for directions on YouTube, we look to the entanglement with others, the *Mind of God*. To me this fits the description of someone considered intuitive. They didn't get there without help from others.

Part VI
Comments

Writing this book seems a bit like saying goodbye to an old and faithful friend. A friend who has provided answers that go beyond one's imagination. I have not shared this friend with others although I have tried. My friend's existence goes beyond a metaphorical nuance. Someone who has traveled such a path can appreciate what I mean.

The Journey

The journey was never about building a preconceived physical structure. It was through Nature's eyes that the building process occurred. First, one had to find the most simple start since it seemed that Nature must be simple in the beginning.

If one allowed magic in the journey there would be no point in searching. Every time the going got tough a person could fall back on magic of some sort. To me this meant that magic fields; or, the wonder of God could not be part of the process.

Disallowing concepts from our physical world placed severe restrictions on the search. One could not use any of our physical characteristics to traverse the foundational thoughts. Herein lies a major problem. How could we ever hope to build a structure from a pile of 'somethings' that had no physical characteristics? In the end I needed to concede that the concept of *neighbor* was okay to use. This did not seem to impart physical characteristics to the 'somethings'. I am not the first person to use the label 'something'. How others mean for it to be used did not influence my thoughts.

A single conglomerate *something* seemed that it would require elaborate relationships in order to be useful. The alternative of multiple somethings required a distinction for when one something ended and another started. To that end I introduced the concept of *nothing* in order to separate the somethings. Today, I view nothing as an imaginary separation that allows useful imaginary concepts to postulate the interaction between neighboring somethings. If this seems to be an impossible starting point, there is good reason. I considered giving up the search at this juncture. Instead, I affirmed my resolve. A few weeks later a way to continue presented itself.

In order to meaningfully explore the concept of events between neighboring elements of space one must know the *laws* by which the elements interact. I failed at attempts to consider this mathematically. After a few years I took the only path that seemed possible for me to consider. At that time I thought *nothing* was empty space much like regular space but absent the fabric. With this idea of nothing I took some manageable directions that continue to surprise me that they worked.

I reasoned that without the magic of fields, a something would be unaware of other somethings. It would also be unaware of any preferred coordinate system. This means that any type of chaotic motion could and would occur. This also means motion in a circle upon a circle, ad infinitum, is possible. Without a reference system the somethings had no way to remain motionless.

I started a math class at the University of Washington, but withdrew from the class largely because I was working nights. I wish I had completed the class. The instructor was phenomenal and the topic was fascinating. I remember a description of Fourier Series that said any periodic motion could be represented by an infinite series of sine and cosine functions. It seemed that whatever Nature did at the foundational level needed to be periodic (i.e. repeatable). It also seemed that circles upon circles could be viewed as sine and cosine functions. The journey continued.

The path out of chaos starts with a random *collision event* between two somethings. The term *collision* comes from motion in an imaginary coordinate system. When two somethings bumped into each other they formed a local reference system, albeit imaginary.

What happens when two somethings collide? I chose the most simple answer. They cancel opposing motions. This is more than a bit simplistic, but it was a manageable thought. Later, it would be found to be seemingly exact. At the foundational level there are no elastic or inelastic collisions. Those types of collisions occur in our physical reality and rely on electromagnetic fields.

It seems that multiple collision events would leave only two main circular motions. The minor circular motions would cancel. This leads to the donut space element and donut chain views.

The journey continues beyond the comments here, but much of it is detail not suitable for general comments.

A Crackpot's Gibberish

You may notice that I avoid the label *Theory of Everything* or TOE. I find the label offensive to the hundreds of thousands of physicists and philosophers who have contributed so much. Their contributions can never be replaced by a Theory of Everything. Claiming such a label may be advertising that one is a crackpot.

Avoiding the TOE label does little to assure the reader that they are reading anything other than the ramblings of a crackpot. It would be hypocritical for me to fault this. I would think the same if I were them. I do not mind if others call me a crackpot. It may mean they actually understood parts of DCT.

The reader may wait for general acceptance of DCT before plunging into it. I expect general acceptance will never happen. I also expect DCT will never be rediscovered. This is not egocentricity speaking. I solved DCT with a great deal of help from others (Section 13.1, Sleep Thinking). I needed their help.

Donut Chain Theory goes far beyond a single-concept theory. It is a journey with many subtle turns. Grasping DCT requires an immersion that few would choose. If a reader wishes to make a serious effort at understanding DCT, I suggest skipping the appendices and chapters flagged as technical. It is vital to grasp entanglement.

DCT could be a sad or bitter discovery if one were seeking fame. My sole motivation was curiosity. DCT has rewarded me immensely. If I place no value on fame then why spend so much time and effort seeking others who might understand? After solving the *ggee ratio* in 2016, I felt an obligation to seek out others who might also be curious. Before getting the *ggee ratio* it seemed unlikely that DCT could be communicated with certainty.

I call this "the book nobody will read". So, why would I write it? I wrote it for two main reasons. First, I wanted my children to have some tangible evidence that my philosophical journey was fruitful. Second, I wanted to document the complete DCT journey in case some future curious person would wish to read it. Both of these reasons required that I write the book in the best fashion I could. It also means I can put DCT on a shelf and live life.

Part VII
Appendices

Calculations have been placed in the appendices in order to facilitate telling the story without the interruption of formulas. The exception to this is Chapter 9, Fine Structure Constant, which is flagged as optional.

Appendix I – *The Egocentricity of Special Relativity* shows how SR can be interpreted to be consistent with having a stationary structure for the foundation. It breaks the SR transformation into two components, real and observational. The observational component accounts for erroneously assuming one is at rest with the underlying structure. Even if the structure cannot be detected, it is important to understand that it can exist.

Appendix II – *Theoretical Ratio of the Gravitational Force to the Electromagnetic Force between Two Electrons (ggee ratio)* provides the calculation of this ratio based on the internal structure of the electron. This calculation seems nearly impossible, but it came directly from Donut Chain Theory; albeit a long journey. The ratio is not forced; nor, does it come from indirectly including the ratio in the input to the calculation. This paper was written when I still believed the imaginary phase geometry in a chain segment behaved similarly to a real geometry. I now believe it simply provides the correct phasing between the elements comprising the chain segment.

The *ggee ratio* demonstrates that DCT correctly describes much of the foundational structure.

Appendix III – *Are Gamma-ray Bursts Caused by Multiverses?* explains GRB's as temporary (burst) connections between separate universes. The explanation seems well fitted to this mystery of the cosmos.

Appendix I

The Egocentricity of Special Relativity

Abstract

This paper assumes the existence of a fabric of space that is locally Euclidean with a preferred coordinate system. These assumptions are shown to produce the special relativity transformations for two bodies in collinear motion. Of primary importance is the insight gained into the transformations of special relativity. There is an observational error factor and an actual component factor for each transformation.

Purpose

This papers allows one to view the fabric of space as being a structure that supports a preferred coordinate system. Special rel ativity is shown to be consistent with this view. The inability to detect when an observer is stationary in the structure of space does not prevent a stationary structure from existing.

Special relativity allows an egocentric view for all observers. This can mislead one into believing special relativity precludes a stationary structure of space. It is important to realize that a stationary structure may be considered when one contemplates the fabric of space.

The Michelson-Morley (M-M) experiment is considered by some to support rejection of an ether based preferred coordinate system consistent with Euclidean geometry. It seems that the M-M null results must occur if the physical apparatus used for the experiment is held together electromagnetically and as such undergoes the same changes as the light waves being examined.

Real Versus Observational Transforms

Assume two observers, A and B, are moving through a preferred coordinate system with velocities u and v. Using the addition rule for relative velocity differences between A and B produces the following :

$$w = \frac{v - u}{1 - u\,v \,/\, c^2} \tag{1}$$

The Special Relativistic transformation using velocity w :

$$\gamma\,(w) = \frac{1}{\sqrt{1 - w^2 \,/\, c^2}} \tag{2}$$

Substituting Equation (1) into Equation (2) and simplifying, gives (hint : square equ. 2 before subst., then sqrt.) :

$$\gamma\,(w) = \frac{c^2 - u\,v}{\sqrt{c^2 - u^2}\ \ \sqrt{c^2 - v^2}} \tag{3}$$

Equation (3) is symmetrical in u and v as expected from a relativistic view. The task remains to determine what part of the special relativity transformation is real and what part is observational. If we adopt the view that a preferred coordinate system exists, it follows that we need to adjust measurements to compensate for our motion in that preferred coordinate system.

If a body is at rest in the assumed preferred coordinate system, then there are no observational errors from the failure to consider motion of the observer. All of the special relativity transformation is real. We can use the inverse of the special relativity transformation to backward reference a body in motion to one without motion. This inversely transformed relationship can then transformed into a different body in motion.

The Lorentz contraction forms the real portion of the transformations between body A and body B. This occurs because a body is held together electromagnetically at some level. Thus, a body and its distance measuring device contract in order to maintain the same electromagnetic equilibrium they had while at rest.

The real component of the transformation of body B observed by body A :

$$\frac{\sqrt{c^2 - u^2}}{\sqrt{c^2 - v^2}} \tag{4}$$

The real component of the transformation of body A observed by body B :

$$\frac{\sqrt{c^2 - v^2}}{\sqrt{c^2 - u^2}} \tag{5}$$

Dividing Equation (3) by Equation (4) gives the observational component of the transformation of body B observed by body A :

$$\frac{c^2 - u\,v}{c^2 - u^2} \tag{6}$$

Dividing Equation (3) by Equation (5) gives the observational component of the transformation of body A observed by body B :

$$\frac{c^2 - u\,v}{c^2 - v^2} \tag{7}$$

Equation (6) or Equation (7) reduces to 1 if u or v, respectively, equals zero.

Combine Equation (4), real factor; and Equation (6), observational factor, giving A's view of B:

$$\begin{array}{ccc} & \text{Real} & \text{Observational} \\ \gamma\,(w) \;=\; & \left[\dfrac{\sqrt{c^2 - u^2}}{\sqrt{c^2 - v^2}}\right] & \left[\dfrac{c^2 - u\,v}{c^2 - u^2}\right] \end{array} \tag{8}$$

or, relativisticly viewed:

$$\gamma\,(w) \;=\; \frac{c^2 - u\,v}{\sqrt{c^2 - u^2}\;\;\sqrt{c^2 - v^2}} \tag{9}$$

Combine Equation (5), real factor; and Equation (7), observational factor, giving B's view of A:

$$\begin{array}{ccc} & \text{Real} & \text{Observational} \\ \gamma\,(w) \;=\; & \left[\dfrac{\sqrt{c^2 - v^2}}{\sqrt{c^2 - u^2}}\right] & \left[\dfrac{c^2 - u\,v}{c^2 - v^2}\right] \end{array} \tag{10}$$

or, relativisticly viewed:

$$\gamma\,(w) \;=\; \frac{c^2 - u\,v}{\sqrt{c^2 - u^2}\;\;\sqrt{c^2 - v^2}} \tag{11}$$

Appendix II

Theoretical Ratio of the Gravitational Force to the Electromagnetic Force between Two Electrons

Abstract

This appendix develops the theoretical ratio of the gravitational force to electromagnetic force between two electrons. I refer to this ratio as the *ggeeratio*. The $ggee_{\text{theory}}$ ratio equals the product of a factor multiplied by $alpha^2$.

The factor portion of the calculation comes from an unusual source. An underlying model posits a metaphysical structure of space and of the electron. A rational number solution to the geometry of the model leads directly to the factor. This solution emerges completely independently from the *ggee* ratio it produces. The rational factor seems to be an exact solution.

The precision of $ggee_{\text{codata}}$ depends on the precision of G. The precision of $ggee_{\text{theory}}$ depends on the precision of $alpha^2$. Using Codata values for 2018, $ggee_{\text{codata}} = 2.400610(54)E-43$ and $ggee_{\text{theory}} = 2.40071068266(72)E-43$. The theoretical value is 1.85 *sigma* greater than the Codata derived value. Based on the value of *alpha* in 1969, $ggee_{\text{theory}} = 2.4007097(72)E-43$. This value is also 1.85 *sigma* greater than the Codata derived value for 2018.

The gravitational constant, G, has a history of disparate value ranges. A deviation of 1.85 *sigma* may fall into an acceptable range more than would normally be the case. The original calculation of ggee was completed March 10, 2016. At that time, the University of Washington had published a paper indicating a *sigma* for G that was about one-third of the 2014 Codata *sigma*. That paper was apparently withdrawn. If the UW result had held that would have meant that this *ggee* theoretical value was a 4 *sigma* departure from the experimental value.

The equations have been reordered and renumbered. A page has been added to show the solution calculation for an electron.

Donuts

In order to follow the *ggee* ratio calculation one needs some understanding of the underlying donut structure. The donut path depicts the dynamics of the structure in a manner that provides both a method to do computations and a suitable visual aid.

The fabric of space under this view derives from *something* and *nothing*. *Something* lacks all physical characteristics. *Something* nondestructively interacts with other *somethings* by simply canceling opposing motion when contact occurs. *Nothing* means the complete absence of anything, even the fabric of space. Concepts of *extension*, *time*, and *motion* lack meaning at this metaphysical level.

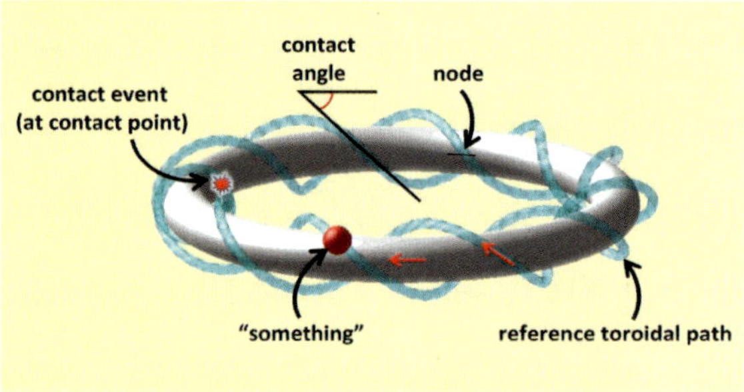

Figure 15.1: Basic donut chain link.

Figure 15.1. shows a dynamic donut path. The donut shown has 11 nodes that complete in two revolutions (because of intertwining). The red ball represents *something*. The remainder of the image serves only to help visualize the path. An animated version of this image may be seen at: http://spaceandmatter.org/dn.htm.

The illustrated donut helps us to visualize the fundamental metaphysical element. The metaphysical level is *event* and *phase* driven. This paper frequently references the number of nodes in a donut. I suggest counting the 11 nodes in Figure 15.1.

All events considered in this paper happen on the inside edge of the donut closest to the donut hole at the toroidal position where adjacent chain links make contact.

Objects normally travel in a straight line in the absence of an external force. The donut motion occurs in *nothing* (a complete void) which allows the circular path.

The donut gains context only by reference to an assumed coordinate system. Donuts periodically contacting adjacent donuts provide the needed context. For donuts to be a stable part of the fabric of space they must be part of a chain segment. The chain segment must be connected to the fabric of space on both ends with the connecting links parallel; or, π radians out of phase.

Chain Segments

Chain segments are the strands used to create both the fabric of space and matter connected to the fabric. Chain segments provide a basis for performing calculations. Chain segments also provide a visualization of the metaphysical structure. It is useful to think of the chain segments as metaphysically real.

What is the difference between *physically real* and *metaphysically real* as they are used in this paper? *Physical reality* comes from that which our senses perceive and our languages reflect. In a physical sense we may feel we understand the meaning attached to distance (extension) and time; or, motion. The meaning of these concepts provides the basis that allows us to form a visual imagery of processes. *Metaphysical reality* may determine our physical reality, but it is removed from our direct scrutiny. This is much like trying to visualize the functional components of a combustion engine by observing its external properties.

Concepts of distance and motion are used in visualizing, understanding and evaluating metaphysical reality. These physical concepts may be considered as artificially introduced in the metaphysical realm. They are helpful and may be necessary for us to understand it. The coordinate system and the donut characteristics may also be considered as chosen to allow us to visualize metaphysical behavior. I personally find it helpful to consider the introduced view of the metaphysical realm to be real at that level. In other words, I find it to be metaphysically real and capable of being considered in a concrete manner.

A chain segment forms from connected donut links. As it is used in this paper, a chain segment connects to other chain segments on both ends. To be a chain segment means that it branches on both ends. We always consider one chain segment as being connected to two other chain segments on each end. This means that each end donut link has a total of three chain segments connected to it.

For chain segment length we count only one of the two connecting end donut links. The orientations of the connecting end donut links are considered to be parallel (either in phase; or, π radians out of phase).

We will later find that the space fabric that we think of as a vacuum develops from chain segments 138 links in length. The electron has one link missing which requires a twist in order to connect. The electron forms from a 137 link chain segment.

Rational Number Relationships

In the metaphysical realm events occur discretely. This requires rational number relationships. Parts of the donut chain solution produce irrational results. In these cases we must find the correct rational representation of the irrational relationship. This assumption yields answers that support its validity.

Contact events drive behaviors in the fabric of space. A contact event between adjacent donuts occurs at a particular location at an instant of time. Each donut has two independent motions for location and one dependent motion for timing. These motions must all synchronize in order to produce stable contact events.

Consider the time elapsed between contact events for two adjacent donuts as taking one big unit of time. This big unit of time needs to be divided into much smaller units in order for each motion to be represented by integer values of elapsed time.

This paper determines each of the three integer values for motion within the electron chain segment. We solve for the toroidal (main circle) and poloidal (outer circle) motion values in the first step of the calculation. We solve for the timing value in the second step of the calculation using values from the first step. The Least Common Multiple of these three integer time values enters directly into the calculation of the *ggee* ratio.

The first major step solves for the donut count in the electron chain segment. This relies on the geometry of the chain segment and understanding how the electron moves through space (i.e., (n-1)(n+1)). The donut model helps handle the mathematics involved. It may be considered as a metaphysically real visual. These relationships produce a clear solution where the toroidal motion has as 76172 units (or $4 \cdot 137 \cdot 139$) and poloidal motion has 74445 units (or $3 \cdot 5 \cdot 7 \cdot 709$).

The first major step synchronizes toroidal and poloidal motion so that the contact point will be revisited. We still need to synchronize the timing. Timing synchronization requires the elapsed time to be an integer value of some timing unit. Elapsed time may be viewed as the hypotenuse of a right triangle where toroidal and poloidal motion units form the triangle sides.

The second major step solves for the rational approximation for the square root of $76172^2 + 74445^2$. We make use of the Least Common Multiple of integer values for the two legs and the hypotenuse (both numerator and denominator). The solution for the hypotenuse is $[(7 \cdot \cancel{7} \cdot 347 \cdot 253153)/(\cancel{3} \cdot 19 \cdot \cancel{709})]$. Factors for the sides and the hypotenuse combined with the electron coupling constant produce the *ggee* ratio.

Event Probabilities

Events forge and closely synchronize the Metaphysical Realm. A synchronized view of contact events provides the means to do calculations in this realm. The donut (i.e., chain link) helps to visualize event synchronization.

Contact between the *somethings* of adjacent donut links constitute a ***contact event***. A contact event can only occur if a ***contact node*** simultaneously occurs in adjacent donuts. This requires the correct *where* and *when* for both donut links.

Visualize a toy Slinky bent completely around on itself to resemble a donut. Think of the Slinky's spiral spring as a path traveled by *something*. Define a ***donut path*** to be this spiral path.

Consider the donut as a link in a chain segment. Chain links make contact with adjacent chain links on the inside of each link. A donut works the same. Define a ***contact line*** as the part of a donut closest to the donut hole. The contact line forms the circumference of a circle that immediately surrounds the donut hole.

A *contact node* occurs at each point where the *donut path* intersects the *contact line*. The contact nodes for a donut constitute the eligible places the next event can occur. The label *"contact node"* refers to a location and instant on the donut path. At that location and instant, contact with another *something* could possibly occur.

Inverting the number of contact nodes yields the frequency with which an event occurs. Complex environments can appear to exhibit a stochastic view of the frequency. Confined environments, such as a particular chain segment, exhibit a deterministic view of the frequency.

A *contact event* between adjacent donuts occurs when their *somethings* arrive simultaneously at connecting contact nodes located in separate donut paths. Several conditions must be met in order for a *contact event* to occur. This may make the events seem unlikely. Remember that nothing happens until a contact event occurs however unlikely the event may seem.

Calculation Overview

The *node calculations* section (equ. 4-8) determines angle relationships used in equation 15.a. The *formula for ggee ratio* section (equ. 9-14) determines the relationship of gravity to electromagnetism using a thought experiment. It assumes that the entire mass of the electron is due to charge. The thought experiment can't actually happen. The geometry of spacetime surrounding the electron has no effect on the relationship. The *calculation of ggee ratio* section (equ. VII-16) determines the ggee ratio using equations 7 and 14. Synchronized rational nodes produce the ratio. For that reason the calculation may hold for the full precision shown.

Node Calculations

Node calculations rely on *contact angles* determined by integer relationships of toroidal and poloidal motion. The donut shown as Figure 15.1. in the Donuts section provides a visualization and labels major donut features.

For a node to exist the donut path traveled must eventually repeat its pattern. Consider the toroidal phase (major circular position) and poloidal phase (minor circular position) of *something* at a moment in the donut path. We choose the donut radii so that the *donut path* angle on the inside of the donut (i.e., at the *contact line*) will exactly produce the desired node count for the donut. They also produce the same motion at the moment of contact with another donut.

Define the connecting angle between donuts as the ***donut contact angle***, $\boldsymbol{\phi}$. The orientation of untwisted chain segment links differs by $\pi/2$ between adjacent links. This produces an angle of $\phi = \pi/4$ connecting two untwisted links. The angle ϕ plays a pivotal role in contact nodes, especially in twisted chain segments.

We describe radii in order to visualize the donuts. The *ggee* calculation *excludes* these radii. Imaginary extension in the metaphysical realm helps us to visualize.

The major (toroidal) radius equals $\boldsymbol{R}$ measured from the center of the donut hole to the inside of the torus (the *contact line*). The major angular velocity equals $\boldsymbol{\Omega}$.

The minor (poloidal) radius equals $\boldsymbol{r}$ measured from the center of the torus ring to the surface of the torus. The minor angular velocity (poloidal) equals $\boldsymbol{\omega}$.

The major angular motion in a donut must equal the minor angular motion in a connecting donut (visualize adjacent space donuts). Thus:

$$\omega r = \Omega R$$

Define $\boldsymbol{p}$ as the number of *primary contact nodes* in one major revolution of a donut link. For a simple untwisted chain segment this produces:

$$\omega = p\Omega \quad \text{and;} \quad pr = R$$

To have nodes at each pth position, there must be some multiple of p poloidal revolutions for each toroidal revolution. The radii relationship provides equal and aligned connected motion between adjacent donuts.

The node calculation process determines solutions based on untwisted (space) chain segments containing $\boldsymbol{n}$ links. The process checks each $\boldsymbol{n}$ within a range of possibilities looking for the best solution.

In order to perform the calculations, we must understand the form of an electron and how it moves through space. This motion leads to a node incremental count. Without this incremental count the calculations provide no useful answers.

The number of links in the electron chain segment is one less than the number of links ($\boldsymbol{n}$) in the untwisted space chain segment. In order for the electron to move it shifts an end connection by one link to restore the electron chain segment to a space chain segment and change the next space chain segment into the electron. If we repeat this process for several electrons following the same path, the connection to space chain segments keeps moving in the opposite direction as the electron. In order to restore the space connection balance the electron path (connected chain segments) needs to move in the opposite direction as the electron moves. The only way it has of doing this is to loop around on itself to have a fully connected moving strand of chain segments.

As the connected strand of chain segments moves through space there momentarily exists $n-1$ and $n+1$ chain segment lengths. A dynamic chain segment containing n links synchronizes best if each link has n nodes. If each link experiences the same shift between links of t nodes then synchronization occurs at each end of the chain segment, because nt equals tn. The incremental node count for electron motion needs to be some multiple of p from the following relationship:

$$p = (n-1)(n+1) \tag{1}$$

Ω equals one. We synchronize on one toroidal revolution. When a solution would take multiple revolutions, we restate the number of nodes, multiplying by m.

$$\Omega = 1 \tag{2}$$

In each donut link what we call *motion* happens along the spiral path. The main toroidal motion and the outer poloidal motion are components of the spiral path motion. Regardless, we choose the elapsed time between the initial toroidal contact node as a fixed reference unit. By using toroidal node spacing for units, we know that integer timing results happen on contact nodes.

$$\Delta\Omega = \frac{1}{mp} \tag{3}$$

When a single link is removed from an untwisted chain segment, the remaining chain segment must be twisted in order to connect. This twist produces a *donut contact angle* different than $\pi/4$. We call this new contact angle $\boldsymbol{\phi}_{\text{target}}$. The target angle is an irrational number which cannot be achieved using rational node counts. Instead, we find a close solution, $\boldsymbol{\phi}_{\text{solution}}$, based on a rational node count.

Consider the twisted chain segment where a link is removed and twist is added. Define the length of the untwisted chain segment prior to removal of a link as equal to $\boldsymbol{n}$. The target donut contact angle for the twisted chain segment equals:

$$\phi_{\text{target}} = \frac{\pi}{4} + \frac{\pi}{2(n-1)} \tag{4}$$

The new angle spreads the twist from the missing link over the remaining $n-1$ links. The target angle solves this relationship exactly, albeit irrational.

For poloidal 'motion' the incremental value of $\Delta\omega_{\text{target}}$ in terms of $\Delta\Omega$ units is non-integer due to the need to achieve the target donut contact angle, ϕ_{target}:

$$\Delta\omega_{\text{target}} = \tan(\phi_{\text{target}}) \cdot \Delta\Omega \tag{5}$$

ω_{solution} equals the rounded (i.e., complete revolutions) inverse of equation (5). This is found by incrementing by $\Delta\omega_{\text{target}}$ until m produces a sufficiently close answer.

$$\omega_{\text{solution}} = \text{round}\left[\frac{1}{\Delta\omega_{\text{target}}}\right] = \text{round}\left[\frac{mp}{\tan(\phi_{\text{target}})}\right] \tag{6}$$

Solve for ϕ_{solution} (based on complete revolutions) using equations (2) and (6):

$$\phi_{\text{solution}} = \tan^{-1}\left[\frac{\omega_{\text{solution}}}{\Omega}\right] \tag{7}$$

The use of tangent differs between equations (5) and (7). Equation (5) relates ideal poloidal nodes to toroidal nodes. Equation (7) expresses the integer twist solution.

Solve for $\phi_{\text{collision}}$ using equations (7) and (4):

$$\phi_{\text{collision}} = \left|\phi_{\text{solution}} - \phi_{\text{target}}\right| \tag{8}$$

Node Instability Measures

We determine m by testing *node instability measures.* These measures are chosen on the basis of judgement. As such, the measures should not be considered inviolable. The instability measure factors used are:

- untwisted chain segment length — n
- collision angle — $\phi^2_{\text{collision}}$
- major toroidal revolution count — m_{solution}
- poloidal revolution count — $\omega^2_{\text{solution}}$

The gross instability measure equals the product of the variables shown (raised to the power shown). Instability measures are normalized to the electron solution which is assigned 1000. The instability measure is intended to represent the relative chance of failing to maintain stability.

Untwisted chain segment length – n

For the untwisted space chain segment we assume the chance of it failing (becoming unstable) is proportional to its length. A longer (more links) space chain segment has more exposure to a destabilizing event.

Collision angle – $\phi^2_{\text{collision}}$

The collision angle directly accounts for the slowing effect from matter. We use the square of this value which coincides with its use in the slowing of time.

Major toroidal revolution count – m_{solution}

The major toroidal revolution count is arbitrarily assumed to contribute instability proportional to the number of major revolutions. This factor may already be accounted for in the poloidal revolution count.

Poloidal revolution count – $\omega^2_{\text{solution}}$

The poloidal revolution count is arbitrarily assumed to contribute instability proportional to the square of its revolutions. Getting the same discrete answer seems less likely as the number of poloidal revolutions increases. Time exposure is increased and there is more opportunity for interacting on the number of nodes.

The same correct solution emerges from a wide range of choices for the instability measures. This renders their choice somewhat academic. For example, if we omit m_{solution} and do not square ω_{solution} we still get the same best answer for the electron.

Electron Structure

The electron structure described in this paper emerges from a metaphysical understanding of how the fabric of space formed from *something* and *nothing*. The logic of the metaphysical underpinnings is not included with this paper.

Normally, objects travel in a straight line in the absence of an external force. The donut motion occurs in *nothing* (a complete void) which allows it to be viewed as traveling in the manner demonstrated.

The fabric of space consists of donut chain segments containing 138 donut links. The electron consists of a donut chain segment 137 links long. The electron chain segment is twisted because it has one fewer donut links than the fabric of space.

Donuts that form the electron chain segment have 74445 nodes that complete in four toroidal revolutions. It takes 76172 toroidal node units to synchronize with the external chain segments of space. Thus, it takes $74445 * 76172$ revolutions to be in the original position. In order to make contact donuts must synchronize time-wise as well as position-wise. This requires a rational value for the resultant vector.

This solution best aligns the angle between donuts that results from having 137 links in the electron chain segment.

The 138 and 137 chain segment lengths for space and the electron, respectively, resulted from solving a relationship. The stability of this solution exceeds the stability of the next best solution by over 2 orders of magnitude. This is the reason dimensionless numbers close to 137 have special significance in physics.

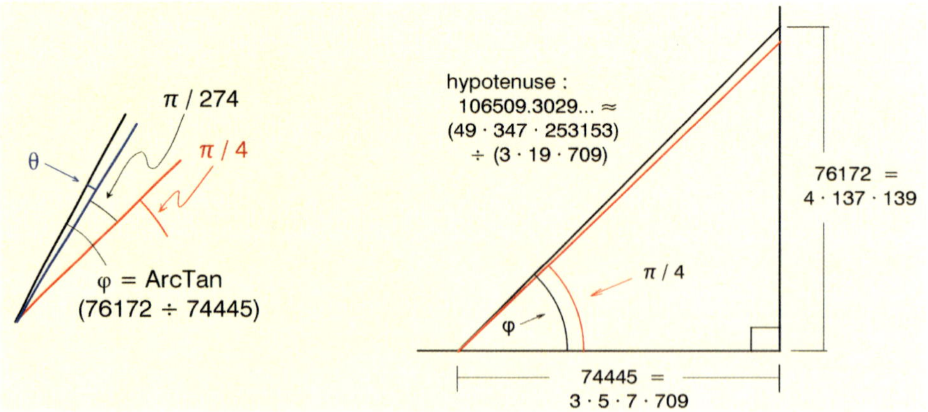

Figure 15.2: Components for the ggee ratio calculation.

Formula for ggee Ratio

The *ggee* ratio calculation originates from the assumption that matter slows time combined with the assumption that the mass of the electron is due entirely to its charge. A concept of drag, d, helps facilitate this analysis. Drag is the slowing of time caused by mass. Drag is considered as being in close proximity to the mass.

The rate of flow of time in close proximity to a mass is slowed by the drag. This allows us to represent the mass as creating a new flow rate of time slowed by the drag, d. For the electron this takes the form:

$$t_{\mathrm{e}} = 1 - d_{\mathrm{e}} \tag{9}$$

If we were to double the mass we double the drag, d. For two electrons this takes the form:

$$t_{\mathrm{e+e}} = 1 - 2d_{\mathrm{e}} \tag{10}$$

For two electrons the flow rate of time is simply multiplied to get the effect of each electron on the other.

$$t_{\mathrm{ee}} = (1 - d_{\mathrm{e}})^2 \tag{11}$$

Expanding:

$$t_{\mathrm{ee}} = 1 - 2d_{\mathrm{e}} + {d_{\mathrm{e}}}^2 \tag{12}$$

Consider Equation (12). The mass associated with two electrons attributed to charge is 2 d_{e}. The mass loss associated with the force of gravity is ${d_{\mathrm{e}}}^2$. This yields the following ratio for *ggee*:

$$ggee = \frac{{d_{\mathrm{e}}}^2}{2d_{\mathrm{e}}} \tag{13}$$

Or:

$$ggee = \frac{d_{\mathrm{e}}}{2} \tag{14}$$

What does Equation (14) mean? We consider the electron structure in answering this question. Later, we will discover that the electromagnetic coupling constant (alpha) squared balances the experimental value with the theoretical value.

The time drag from the electron, d_{e}, can be described by considering the relationship between a normal space chain segment and an electron chain segment. The normal space chain segment interacts with adjacent space chain segments at a regular interval. This is the interval of time over which we measure d_{e}. The electron chain segment is the unit of space that generates d_{e}.

Calculation of ggee Ratio

The theoretical drag, d_{e}, in Equation (14) separates into component factors:

$$d_{\mathrm{e}} = \underset{(15.a)}{\begin{bmatrix}\text{motion lost}\\ \text{for each}\\ \text{contact event}\end{bmatrix}} \underset{(15.b)}{\begin{bmatrix}\text{number of}\\ \text{internal}\\ \text{contact events}\end{bmatrix}} \underset{(15.c)}{\begin{bmatrix}\text{frequency of}\\ \text{internal}\\ \text{contact events}\end{bmatrix}} \underset{(15.d)}{\begin{bmatrix}\text{frequency of}\\ \text{external}\\ \text{contact events}\end{bmatrix}}$$

Motion lost for each contact event, using equation (7):

$$1 - \cos(\phi_{\mathrm{collision}}) \approx \frac{\phi^2_{\mathrm{collision}}}{2} = \frac{\left[\tan^{-1}\left(\frac{76172}{74445}\right) - \frac{\pi}{4} - \frac{\pi}{274}\right]^2}{2}$$

$$= \frac{[8.08727858986336E - 11]^2}{2} \tag{15.a}$$

Number of internal contact events (two for each connection):

$$2n = 2 \cdot 137 \tag{15.b}$$

Frequency of internal contact events with both the numerator and denominator of the hypotenuse as factors (cancellations from using Least Common Multiple), equation 15.c:

$$freq_{\mathrm{int}} = \left[\frac{1}{3 \cdot 5 \cdot 7 \cdot 709}\right]\left[\frac{1}{4 \cdot 137 \cdot 139}\right]\left[\frac{1}{7 \cdot \cancel{7} \cdot 347 \cdot 253153}\right]\left[\frac{1}{\cancel{3} \cdot 19 \cdot \cancel{709}}\right]$$

Frequency of external contact events from 2 ends with 3 nodes in the connecting link and square of the electron coupling constant (one for each electron):

$$freq_{\mathrm{ext}} = \frac{2}{3}\alpha^2 \tag{15.d}$$

Ratio of gravitational force to electromagnetic force between two electrons substituting equation (VII) into equation (14):

$$ggee = \frac{1}{2} \cdot [15.a] \cdot [15.b] \cdot [15.c] \cdot [15.d]$$

or,

$$ggee = \left[4.50826219213487 \times 10^{-39}\right] \alpha^2 \tag{16}$$

This may be an exact relationship with precision dependent only on the computational precision and the precision of the electron coupling constant.

Chronological Comparisons of ggee Ratios

The constants below are used to calculate $ggee_{\text{codata}}$ using the relationship $ggee_{\text{codata}} = \mathrm{G} \cdot m_{\mathrm{e}} / \mathrm{c}^2 / r_{\mathrm{e}}$.

	Codata physical constants used in $ggee_{\text{codata}}$ calculation[2] [3] [4]			
	G $[m^3 kg^{-1} s^{-2}]$	m_{e} $[kg]$	c $[ms^{-1}]$	r_{e} $[m]$
×	10^{-11}	10^{-31}	10^{8}	10^{-15}
1969	6.6732(31)	9.109 558(54)	2.997 925 00(100)	2.817 939(13)
1973	6.6720(41)	9.109 534(47)	2.997 924 58(1.2)	2.817 9380(70)
1986	6.672 59(85)	9.109 3897(54)	2.997 924 58	2.817 940 92(38)
1998	6.6730(10)[5]	9.109 381 88(72)	2.997 924 58	2.817 940 285(31)
2002	6.6742(10)	9.109 3826(16)	2.997 924 58	2.817 940 325(28)
2006	6.674 28(67)	9.109 382 15(45)	2.997 924 58	2.817 940 2894(58)
2010	6.673 84(80)	9.109 382 91(40)	2.997 924 58	2.817 940 3267(27)
2014	6.674 08(31)	9.109 383 56(11)	2.997 924 58	2.817 940 3227(19)
2018	6.674 30(15)	9.109 383 7015(28)	2.997 924 58	2.817 940 3262(13)
2022	6.674 30(15)	9.109 383 7139(28)	2.997 924 58	2.817 940 3205(13)

Table 15.1: Chronology of Selected Codata Physical Constants

	calculated	**theory calculated from α_{codata}**		**ratio**
	$gggee_{codata}$	α_{codata} [2] [3] [4]	$ggee_{theory}$	$\frac{ggee_{theory}}{ggee_{codata}}$
×	10^{-43}	10^{-3}	10^{-43}	1
1969	2.4003(11)	7.297 351(11)	2.400 7097(72)	1.00017(46)
1973	2.3998(15)	7.297 3506(60)	2.400 7094(39)	1.00038(61)
1986	2.400 00(31)	7.297 353 08(33)	2.400 711 02(22)	1.00030(13)
1998	2.400 14(36)[5]	7.297 352 533(27)	2.400 710 659(18)	1.00025(15)
2002	2.400 57(36)	7.297 352 568(24)	2.400 710 682(16)	1.00006(15)
2006	2.400 60(24)	7.297 352 5376(50)	2.400 710 6618(33)	1.00005(10)
2010	2.400 44(29)	7.297 352 5698(24)	2.400 710 6830(16)	1.00011(12)
2014	2.400 53(11)	7.297 352 5664(17)	2.400 710 6807(11)	1.000075(46)
2018	2.400 610(54)	7.297 352 5693(11)[6]	2.400 710 682 66(72)	1.000042(22)
2022	2.400 610(54)	7.297 352 5643(11)	2.400 710 679 37(72)	1.000042(22)

Table 15.2: Codata Calculated versus Theory Calculated values for *ggee*

Table 15.1. values and Table 15.2. G_{codata} values come directly from legacy fundamental value tables. The *ggee* ratio values result from substituting α into equation (16).

The constant included in equation (16) is exact and does not vary. The identical constant would have emerged in 1969 had the theory been completed at that time. It is important to realize that equation (16) comes directly from theory.

The results in Table 15.2. may lead one to believe the theory's greatest value lies in the greatly improved precision for the gravitational constant. It does not. The precision of equation (16) and Table 15.2. validate the underlying metaphysical assumptions about the nature of the universe. This understanding provides the greatest value.

Table 15.2. does not provide backward validation of equation (16). Rather, Table 15.2. provides a perspective for the precision of historical values for r_e, m_e and G_{codata}.

The *ratio* column of Table 15.2. indicates a bias (all ratios exceed unity). The source of this possible bias has not been determined.

References

2. B. N. Taylor, W. H. Parker, and D. N. Langenberg, Rev. Mod. Phys. **41**(3), 375-496 (1969)
3. E. R. Cohen and B. N. Taylor, J. Phys. Chem. Ref. Data **2**(4) 663-734 (1973)
4. CODATA Recommended Values of the Fundamental Physical Constants 1986, 1998, 2002, 2006, 2010, 2014, 2018, 2022. https://physics.nist.gov/cuu/Constants/
5. The 1998 Codata value for G is assumed to be 6.6730(10)... rather than 6.673(10)... as stated in the Codata constants listing.
6. The 2018 Codata value for *alpha* appears to be corrected in 2022.

Solution Values for the Electron

This section provides the solution calculations to equations 1-8 for the electron. The precision has been limited for illustration purposes.

$$\begin{aligned} n &= 138 \quad \text{(from solution input)} \\ m &= 4 \quad \text{(from solution output)} \end{aligned}$$

$$p = (n-1)(n+1) \; = \; 137 \cdot 139 \; = \; 19043 \tag{1}$$

$$\Omega \; = \; 1 \tag{2}$$

$$\Delta\Omega \; = \; \frac{1}{mp} \; = \; \frac{1}{4 \cdot 19043} \; = \; \frac{1}{76172} \tag{3}$$

$$\phi_{\text{target}} \; = \; \frac{\pi}{4} + \frac{\pi}{2 \cdot (n-1)} \; = \; \frac{\pi}{4} + \frac{\pi}{274} \; = \; 0.7968638300163892 \tag{4}$$

$$\Delta\omega_{\text{target}} \; = \; \tan(\phi_{\text{target}}) \cdot \Delta\Omega \; = \; 1.34327355744269\,\text{E-5} \tag{5}$$

$$\omega_{\text{solution}} \; = \; \text{round}\left[\frac{1}{\Delta\omega_{\text{target}}}\right] \; = \; \text{round}\,[74445.000012044] \; = \; 74445 \tag{6}$$

$$\phi_{\text{solution}} \; = \; \tan^{-1}\left[\frac{\omega_{\text{solution}}}{\Omega}\right] \; = \; \tan^{-1}\left[\frac{76172}{74445}\right] \; = \; 0.7968638300972620 \tag{7}$$

$$\begin{aligned} \phi_{\text{collision}} \; &= \; \left|\phi_{\text{solution}} - \phi_{\text{target}}\right| \\ &= \; \left|0.7968638300163892 - 0.7968638300972620\right| \\ &= \; 8.08728\,\text{E-11} \end{aligned} \tag{8}$$

Appendix III

Are Gamma-ray Bursts Caused by Multiverses?

Abstract

Multiverses provide a causal mechanism for gamma-ray bursts(GRBs). Assume that differing clock rates prevent interaction between universes in a multiverse. Gravitational time dilation in one universe may allow a temporarily connection to a slower universe. The formation and breaking of such a connection would produce neutrino emissions, gamma-ray emissions, and after-glow. This view derives from looking for a candidate that could connect universes; rather, than looking for an explanation of GRBs. email: rlmarker@spaceandmatter.org

Connect/Disconnect Process

What might we expect if time dilation in a black hole(BH) allows a faster universe to synchronize and connect with a slower universe? Energy and matter from the BH would be attracted into the slower universe. The energy and matter transfer slows the slower universe and speeds up the originating universe in the region of the connection. The clock-rate changes desynchronize and separate the connected universes. The process repeats itself as each universe regains an equilibrium.

Connecting

Connecting the space fabrics of two universes releases neutrinos from duplicate connections and produces gamma-rays from motion between the universes.

In the region of the connection, strands of the fabric of space would be duplicated. Duplicate strands disassociate from the fabric to produce neutrinos. Relative motion, especially circular motion, between the two universes would appear as magnetism or electromagnetic waves. Energetic particles would also be created.

The views in this paper originate from a reasoned and compelling way of looking at the structure of space and matter[1].

Disconnecting

Disconnecting joined universes would leave vacancies in their space fabrics due to the loss of neutrinos during the earlier connection process. Some of the matter in the faster universe may also have moved into the slower universe.

A candidate for repopulating the vacancies would be a building process similar to the building process that *grows* a universe at its boundary. Comparing the GRBs' initial rebuilding glow with the cosmic microwave background(CMB) may provide clues about the nature of the GRBs' afterglow. It may also suggest an alternative explanation for the Big Bang and CMB.

GRBs from Binary Neutron Star or BH Merger

Could GRBs in one universe originate from black holes in a different universe? Could the recipient universe possibly gain a portion of the BH during this process? Janiuk et al (2017)[2] propose events that suggest causal relationships may exist between BHs and GRBs.

Janiuk's paper explores events from the relationships of physics. This paper makes no attempt to provide a physical analysis of the novel view presented. With the temporary conjunction of two universes, usual conservation rules do not apply.

Binary Neutron Star Merger – Late Time Afterglow

Lazzatid et al (2018)[3] observe a late time increasingly luminous afterglow from BNS merger GW170817. A faint gamma-ray pulse almost proved that BNS mergers are associated with some short GRBs, albeit not the canonical short GRBs which exhibit much greater luminosity.

With the multiverse view of GRBs, the luminous burst and release of neutrinos would result when a connection formed between two universes. When the universes separate from each other a repair to the fabric of space would follow. The repair process would be the very process that *grows* the universe and may produce the CMB.

The repair process happens over time and in a somewhat random fashion. Many connections would go through a process that creates matter and energy at the same time the fabric repairs itself.

Why Multiple Universes?

The thoughts that prompt the multiverse view make a difference. If the multiverse thought lacks a foundation, it carries no real value as an explanation for GRBs. The space structure underlying these thoughts suggests the likelihood of many separate universes. To be separate, a universe must have a different clock rate or handedness. This should not be confused with many-worlds, parallel universes, or extra dimensions.

The idea that the temporary connection of separate universes may actually be a GRB came from looking for possible signs of such a connection. The GRBs possess characteristics that make them a suitable candidate to consider.

The concept of independent universes originates from an unpublished metaphysical derivation of the structure of space and matter. Extension (ability to measure distance) lacks meaning in this derived structure. Regardless, we view the derived structure as having extension for visualization purposes. The visualized structure yields a direct calculation of the ratio of the gravitational force to electromagnetic force between two electrons[4].

In addition to equal clock rates, several other conditions need to be satisfied for two universes to have their local space synchronized. This paper assumes those conditions are satisfied. If the conditions always fail to be satisfied, then this paper may be rendered moot.

References

1. Twisted Donut Chain Model of Space, Matter and Origin of Gravity. c1996-2018. Mount Vernon (WA): R. Marker. http://spaceandmatter.org, Section 26. New Model of the Universe.
2. A. Janiuk, et. al. (2017, January 26). Black hole accretion in gamma ray bursts. arXiv:1701.07753[astro-ph.HE]
3. D. Lazzatid, et. al. (2018, May 11). Late time afterglow observations reveal a collimated relativistic jet in the ejecta of the binary neutron star merger GW170817. arXiv:1712.03237v4[astro-ph.HE]
4. R. Marker (2019, September 21) Theoretical Value for Gravitational Constant. viXra:1909.0463, 14-15. [See Appendix II.]